Dust in Galaxies

By

David A. Williams
University College London, UK
Email: daw@star.ucl.ac.uk

and

Cesare Cecchi-Pestellini
Osservatorio Astronomico di Palermo, Italy
Email: cesare.cecchipestellini@inaf.it

ROYAL SOCIETY
OF **CHEMISTRY**

Print ISBN: 978-1-78801-505-9
EPUB ISBN: 978-1-78801-925-5

A catalogue record for this book is available from the British Library

The Royal Society of Chemistry is a charity, registered in England and Wales, Number 207890, and a company incorporated in England by Royal Charter (Registered No. RC000524), registered office: Burlington House, Piccadilly, London, W1J 0BA, UK, Telephone: +44 (0) 20 7437 8656.

Visit our website at www.rsc.org/books

Printed in the United Kingdom by CPI Group (UK) Ltd, Croydon, CR0 4YY, UK

Dust in Galaxies

Preface

At a casual glance with naked eye, the space between the stars – interstellar space – appears to be empty. However, more than two centuries ago the suggestion was made that there are regions of space where starlight seems to be extinguished, and this was the first indication that something might be present in interstellar space: this space might not be empty. At first, the regions where extinction of starlight occurred appeared to be quite well-defined "dark clouds" in which stars could only be seen with difficulty, but evidence gradually accumulated of a general absorption of starlight throughout all of interstellar space. It became clear that the light from distant objects, such as stars, clusters of stars, and galaxies appeared weaker than expected because something was absorbing some of the light along its path to Earth.

What could these interstellar absorbers be? The amount of absorption was found to vary with the colour of the light in a way that could be explained if space contained a population of small dust grains, with sizes comparable to the wavelength of light. Extending observations to longer wavelengths (into the infrared) and shorter wavelengths (into the ultraviolet) showed that dust grains of a wide range of sizes exist in space. This dust is mixed with an interstellar gas – mostly hydrogen atoms and molecules – that had been discovered in the mid-20th century by telescopes

Dust in Galaxies
By David A. Williams and Cesare Cecchi-Pestellini
© David A. Williams and Cesare Cecchi-Pestellini 2020
Published by the Royal Society of Chemistry, www.rsc.org

operating in the radio and ultraviolet regimes. We now think of this interstellar gas and dust as a reservoir of material from which new stars and planets can form.

Up to the mid-20th century, many astronomers were interested in the evolution of stars and galaxies, and dust grains were considered simply an unfortunate impediment to making precise observations of distant objects. So, astronomers found ways of allowing for the effects of dust grains on their observations. However, other astronomers became interested in the dust grains themselves. What are these grains made of and where do they come from? What happens to dust grains in space? Are they merely playing a passive role in the interstellar medium by weakening starlight, or could they be active participants in processes in the Milky Way and other galaxies?

The last half-century has seen a complete revolution in our thinking about interstellar dust. Far from being an unwanted constituent of interstellar gas, merely impeding the observations of stars and galaxies by creating an interstellar fog, astronomers have come to realise that dust grains in the Milky Way have had a number of absolutely fundamental functions in our galaxy's evolution. There is no doubt that our galaxy and others would be very different without the presence of dust. In particular, of course, without dust we would not have a planet for our home, and we would not be here to observe our galaxy.

The role of dust grains in providing molecules that may be involved in the origin of life and in the safe transmission of those molecules to newly-forming planets orbiting Sun-like stars has been one of the greatest surprises of all. Yet such processes now seem to be a natural consequence of all the roles that dust grains play in the interstellar medium. The story of dust in the modern Universe is one that we have learned bit by bit, chapter by chapter, and it is now necessary to consider the impact of the story as a whole.

Our intention in this book is to present the story of dust in galaxies like the Milky Way. Of course, this story is still evolving and there are many uncertainties remaining. We shall tell this unfinished story as best we can in general terms, without mathematics, and using simple chemical ideas. We begin with a discussion of the kinds of observations that led to the idea that dust exists in space, and use this to infer the types of materials of which dust is composed. Very crudely, these materials can be thought of as soot and sand, and the grain sizes are roughly between one thousandth and one hundredth of the thickness of a human hair. Even smaller 'grains' exist, but these should be considered as molecules rather than bulk materials. We then describe the origin of dust in stellar explosions and winds from old stars, and we explore the way that dust evolves in the interstellar medium where it is exposed to starlight, to fast particles known as *cosmic rays*, to high-speed shocks from supernovae, and to other dramatic events. The interstellar medium is indeed a very violent place, and dust grains are modified significantly during their time in interstellar space.

A large part of the book is devoted to the main roles of dust in star and planet formation, and in chemistry that occurs to create molecules involved in the origin of life. These roles begin with the development of interstellar chemistry based on molecular hydrogen – itself formed on dust – in which reactions in the gas with molecular hydrogen are important. The next stage is the formation of simple ices that are deposited on the surfaces of dust grains in denser regions of interstellar space. These ices are particularly relevant for the formation of complex molecules involved in chemistry that occurs before the origin of life. During star formation, dust grains are drawn into a circumstellar disc within which planets can form. Therefore, it seems probable that complex products of the chemistry of simple ices can be safely delivered to a newly-forming planet.

Some of the material in this book is arranged into "boxes" containing additional information for the interested reader, but these can be ignored without affecting the narrative. The book is provided with many images and diagrams.

The book is aimed primarily at an interested reader who comes to the subject without any special knowledge (although a basic background in chemistry may help). The book may also be useful as a course-book for a non-mathematical entry-level lecture course on astronomy and the origin of life.

David A. Williams and Cesare Cecchi-Pestellini

Dedication

We dedicate this book to the memory of Theodore Stecher (1930–2017) whose astronomical observations in 1965 indicated the presence of carbon in interstellar dust grains. This discovery opened up the possibility of complex chemistry in space. We now know that interstellar carbon chemistry is abundant.

Dust in Galaxies
By David A. Williams and Cesare Cecchi-Pestellini
© David A. Williams and Cesare Cecchi-Pestellini 2020
Published by the Royal Society of Chemistry, www.rsc.org

Contents

Dust in Galaxies
By David A. Williams and Cesare Cecchi-Pestellini
© David A. Williams and Cesare Cecchi-Pestellini 2020
Published by the Royal Society of Chemistry, www.rsc.org

CHAPTER 1

Interstellar Dust in Galaxies

1.1 DIRT IS GOOD

We live in a dirty galaxy. It's one of billions of galaxies in the Universe. Almost all of them are dirty. Our galaxy, the *Milky Way*, is dirty because it contains a lot of dust, very roughly about a billion solar masses of dust. That's a lot of dust, because a solar mass (*i.e.*, the mass of the Sun) is huge – about two million billion trillion tonnes. This dust is found to be mixed with a gas that occupies the space between the stars. In total, this interstellar gas is a hundred times more massive than the dust, and is almost entirely hydrogen. The gas and dust together are called the *interstellar medium*, which makes up about one tenth of the total mass (*i.e.*, stars plus gas and dust) of the Milky Way. Astronomers tend to measure masses in terms of solar masses, because the units of tonnes or kilogrammes that we use on Earth seem inappropriate for extra-terrestrial space. Using the mass of the Sun as the unit of mass, then the entire mass of the Milky Way galaxy, including stars, gas and dust is roughly one trillion (or 10^{12}) solar masses; the mass of the interstellar medium (gas and dust) of the Milky Way is about one hundred billion (or 10^{11}) solar masses; and the mass of the dust in the Milky Way is approximately one billion (or 10^9) solar masses.

Dust in Galaxies
By David A. Williams and Cesare Cecchi-Pestellini
© David A. Williams and Cesare Cecchi-Pestellini 2020
Published by the Royal Society of Chemistry, www.rsc.org

In all galaxies, the interstellar medium is the source of matter – mainly hydrogen – from which new stars are formed. The Milky Way is thought to be forming new stars at an average rate of about one star per year. Interstellar space is also the place into which dying stars eject their debris, a lot of which is dust. So, the interstellar medium of a galaxy tends to get dirtier and dirtier (more and more dusty) as the galaxy becomes older. New stars replace older stars that have used up their fuel (hydrogen) and which eventually die; for example, our Sun is middle-aged and will begin to run out of fuel, go through some exciting death throes, and die in about five billion years (long enough so that we don't need to worry). As we shall see in later chapters, interstellar dust plays crucial roles in the processes of creating new stars from interstellar gas, new planets in orbit around these stars, and new comets and meteorites traversing these planetary systems. Some of these planets may be suitable for life, and dust plays an important part in creating molecules that may be involved in biochemistry and the initiation of life itself on new planets (or, some say, even in space). Studies related to extraterrestrial life are very new topics of research, and are called *astrobiology*.

We'll argue in this book that interstellar dust in the Milky Way and in many other galaxies in the Universe is an essential ingredient in the recipe for forming stars and planets, and in the provision of molecules that are necessary for life to begin. We use the word "essential" because – from the biased point of view of human beings on planet Earth – we simply wouldn't exist were it not for the roles of dust. A "clean" Universe is not a place where life as we know it could evolve. It is good to be dirty!

In this book, we shall describe how we know that dust is present in the interstellar medium of the Milky Way (Chapters 1, 2, and 3), and what we know of the "life cycle" of dust from its formation in stars to its destruction in space (Chapters 4 and 5). We shall investigate the physical and chemical properties of dust that allow it to play characteristic roles in the interstellar medium, and we shall explore these roles in Chapters 6–10. We shall come to the conclusion that a galaxy without dust is not a place in which life would be expected to exist. Indeed, the existence of life appears to depend on the presence of dust.

In the remainder of this introductory chapter, we shall set the scene by describing the structure of the Milky Way and its interstellar medium. Although we shall talk frequently about dust in the Milky Way, our discussion will apply generally to almost all galaxies. We'll review a little of the history of the discovery of dust in the Milky Way, and we'll remind ourselves that while we must consider that dust in space is a good thing, dust here on Earth can be rather bad for us.

1.2   THE MILKY WAY AND OTHER GALAXIES

The Milky Way is very big. Imagine a disc of diameter 100 000 light years and thickness 1000 light years. A *light year* is the *distance* travelled in one year at the speed of light, which is about 300 000 km s^{-1}, and so one light year is roughly ten thousand billion kilometres, or 10^{13} km. This imagined disc encompasses such a very large volume that the even the huge amount of interstellar matter in the Milky Way has a very low density indeed. In fact, the *average number density of the interstellar gas* is about one hydrogen atom per cubic centimetre. That average number density – one atom per cubic centimetre – is very low indeed compared to the number density of molecules (oxygen and nitrogen) in the air that we breathe in Earth's atmosphere. The average number density of dust grains in the Milky Way is very much lower than the gas density. On average, there is about one large dust grain in a cube of interstellar space of side 100 metres (m). This is incredibly clean, absolutely immaculate by terrestrial standards! But there's a lot of space in the galaxy, and so there's a lot of dust, too.

However, the interstellar medium in the Milky Way and other galaxies – the gas plus dust – isn't distributed smoothly, but is very clumpy. The density may be many thousands of times larger than the average in some small locations, while large regions may have densities that are very much lower than the average (we'll return to this in the following section). Dust and gas are generally well mixed within galaxies, so that where the gas is denser, it contains proportionally more dust.

We can't step outside our Milky Way to view its shape and size, but we can look at other galaxies, some of which we think are similar to the Milky Way. Figure 1.1 (left) shows an optical image of the Andromeda galaxy, taken in visible light (*i.e.*, light to which our eyes respond). Andromeda is a *spiral galaxy* believed to be similar to the Milky Way. However, the Andromeda galaxy is thought to contain about twice as many stars and be about twice as massive as the Milky Way. Andromeda is the nearest major galaxy to the Milky Way, and is about 2.5 million light years away. It is approaching the Milky Way at about 110 km s^{-1} and will eventually collide with it. That would be a collision worth watching, but we will have to wait for some billions of years before it happens.

The optical image, Figure 1.1 (left), clearly shows the *spiral* structure of the Andromeda galaxy. The dark 'lanes' trace denser regions of the interstellar medium, where dust mixed with the interstellar gas absorbs so much of the starlight that these regions

Figure 1.1   The Andromeda galaxy, (left) optical image (Reproduced from https://commons.wikimedia.org/w/index.php?curid=12654493 under the terms of the CC BY 2.0 license, https://creativecommons.org/licenses/by/2.0/deed.en) and (right) infrared image (credit: ESA/Herschel). In the optical image, the spiral structure of the galaxy is traced in the white light of bright stars that are too far from Earth to be resolved. In some places, the starlight is obscured by clouds of dusty gas, and these clouds appear dark in the image. However, when dust absorbs starlight it is heated and then it radiates in the infrared. The right image shows the Andromeda galaxy in emitted infrared radiation at a wavelength of 24 micrometres (or 10^{-6} metres, or μm). The dark region in the optical image coincides with the bright regions of the infrared image.

appear dark. The energy of the starlight that is absorbed by the dust makes the dust warm enough for it to radiate in the infrared part of the spectrum (see Box 1.1 if you would like more information about the spectrum of radiation). Figure 1.1 (right) shows an image of the Andromeda galaxy taken not in visible light but in this infrared radiation. By comparing the two figures, we can see

---

Box 1.1   Basic ideas of radiation

Light is part of the electromagnetic spectrum. It is a waveform in combined electric and magnetic fields (called the *electromagnetic field*) that travels at a speed c of about 300 000 km s^{-1}. We generally use the word "light" to refer to radiation to which our eyes respond. When a beam of white light (say, from the Sun) is passed through a glass prism, the beam is split into the familiar rainbow of visible colours between red and violet. These different colours of light that our eyes detect are in fact simply different wavelengths of electromagnetic waves. Red light has wavelengths of about 700 nm (1 nm or "nanometre" is one billionth of a metre) and violet light has wavelengths of about 400 nm. Human eyes evolved naturally to detect light as the familiar spectrum of red, orange, yellow, green, blue, indigo, and violet between these limits, but not outside these limits because sunlight is most intense between the limits of these wavelengths. It's amusing to consider that if the Sun were hotter, human eyes would have evolved a sensitivity at wavelengths shorter than violet (*i.e.*, the ultraviolet) and if the Sun had been cooler we would have eyes sensitive to wavelengths longer than red (the infrared).

The electromagnetic spectrum extends far beyond red wavelengths to much longer wavelengths, and to much shorter wavelengths than those of violet. These radiations are just as real as the light we can see, but our eyes don't respond to wavelengths outside the visible range. However, we can build instruments that detect or emit non-visible radiation. At wavelengths of increasing size beyond visible red light there is infrared radiation, microwave radiation and radio radiation. Wavelengths shorter than violet light of decreasing size correspond to ultraviolet (UV) radiation, X-rays, and γ-rays.

If we could observe one electromagnetic wave, we would see the waveform passing us by at speed c, and we could count the number of wave peaks (each separated by one wavelength, λ) passing in one second. This number is called the *frequency*, ν. Obviously, the speed of the wave is equal to the number of wave peaks, each separated by one wavelength, passing by in one second, so the speed c is simply the product of the wavelength, λ, and the frequency, ν.

For example, taking the speed of light to be 300 000 k s^{-1}, ultraviolet radiation with a wavelength λ of 300 nanometres (or 10^{-9} metres, or nm) has a frequency ν of about 1 million billion per second, or in more convenient notation, 10^{15} Hertz.

Radiation packs an energy punch (called a *photon*) that is proportional to frequency ν. Thus, radio waves, which have relatively long waves and relatively low frequencies, have low-energy photons, while X-rays – with short wavelengths and high frequencies – have high-energy photons.

Figure 1.2 illustrates the electromagnetic spectrum, showing the very small visible region of the entire spectrum.

(*continued*)

Box 1.3   *(continued)*

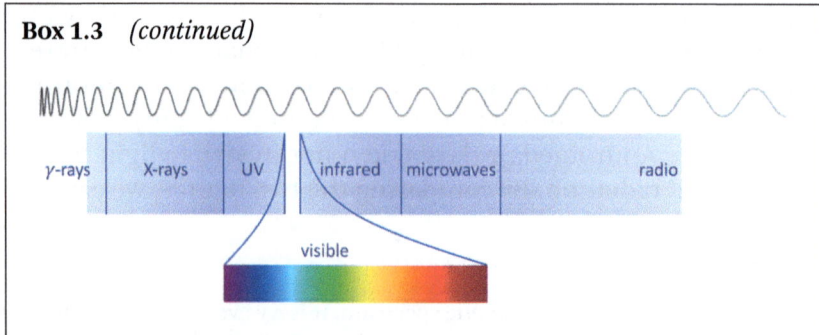

Figure 1.2   The electromagnetic spectrum. The figure shows the electromagnetic spectrum, ranging from γ-rays at the shortest wavelengths to radio waves at the longest wavelengths. The visible region, ranging from violet to red, is a tiny part of the entire spectrum.

that this emitted infrared radiation arises in the places where the optical starlight is absorbed; *i.e.*, the infrared radiation is coming from warm dust.

The Milky Way and Andromeda are both examples of spiral galaxies, but galaxies come in many shapes and sizes (some examples are shown in Figure 1.3). Some have relatively little interstellar gas and dust, and their period of active star formation is therefore drawing to a close since the reservoir of matter for new stars – the interstellar gas – is nearly empty. Unlike spirals, such galaxies may show a smooth near-ellipsoidal structure, and are called *elliptical galaxies*. Other galaxies have large reservoirs of gas and dust, and consequently they may be forming stars at a high rate, turning interstellar matter into stars; these are called *starburst galaxies*. Galaxies do not always show the elegant symmetries possessed by spirals or ellipticals, and some have quite messy shapes; such galaxies are called *irregular galaxies*. Galaxies in the process of formation are called *proto-galaxies*, and probably form from the accumulation of smaller objects. All galaxies rotate, and every part of every galaxy is spinning to a greater or lesser extent. The familiar spin of the Earth that defines our day length, and its orbit around the Sun that defines our year, are examples of the spin that arises ultimately from the spin of the Milky Way.

Galaxies have a wide range of masses. Those with small masses are intrinsically faint, while more massive galaxies tend to be

(a) (b)

(c) (d)

(e) (f)

Figure 1.3   Some examples of images of various types of galaxy. (a) The Pinwheel galaxy (also called M101) is a spiral galaxy, viewed almost face-on from Earth. It is about 21 million light years away. The spiral structure is very clear with the galaxy in this orientation. M 101 has about ten times as many stars as the Milky Way, and its diameter is almost twice as great (credit: NASA, ESA and

intrinsically bright. Of course, bright galaxies that are at very great distances appear faint, while faint galaxies that are relatively close may appear bright. Some basic information about a few galaxy types is given in Table 1.1, and to help us to get used to the distances involved in interstellar and intergalactic spaces, we show in Table 1.2 some relevant or typical astronomical distances.

Table 1.2 indicates the need to use an appropriate unit of distance. Kilometres or thousands of kilometres are certainly handy units for distances on Earth, but are cumbersome units even for distances within the Solar System; the distance to the Sun is so great that sunlight that we see on Earth has been travelling for over eight minutes to reach us. Outside the Solar System the nearest stars are a few light years away. Distances within the Milky Way are usefully measured in thousands of light years. Distances between galaxies can be appropriately

---

the Hubble Space Telescope). (b) The galaxy ESO 325-G004 is an example of a giant elliptical galaxy. Elliptical galaxies have an ellipsoidal shape and a rather smooth profile (none of the stars in this galaxy can be resolved). Ellipticals are dominated by low-mass stars and have few high-mass bright stars. They have little interstellar gas and so star formation only occurs at a low rate compared to the Milky Way star formation rate of about one star per year. ESO 325-G004 is a large elliptical, and is about as massive as the Milky Way. It is about 450 million light years away (credit: NASA, ESA and The Hubble Heritage team and J. Blakeslee). (c) The galaxy NGC 1427A is an example of an irregular galaxy. It is 52 million light years away. Irregulars are often small, and their shapes are affected by near-collisions with more massive galaxies (credit: NASA, ESA and The Hubble Heritage Team). (d) A starburst galaxy is notable for its very high rate of star formation, which may be a hundred times greater than the star formation rate in the Milky Way. The high activity can be stimulated by a collision between two galaxies. The Antennae Galaxies are an example of such a collision. It is a merger between galaxies NGC 4038 and NGC 4039 (credit: ESA/Hubble & NASA). (e) The galaxy IRAS 14348-1447 is an example of an Ultraluminous Infrared Galaxy (ULIRG). Such galaxies emit more radiation in the infrared than at all other wavelengths, and are at least ten times more luminous than the Milky Way. These ULIRGs are very bright because they have very high star formation rates. This particular example is about one billion light years away from Earth (credit: ESA/Hubble & NASA). (f) The image is of a massive object called Himiko that is very far indeed from Earth, about 12.9 billion light years away. It appears to be a galaxy in the process of formation from three objects. It is called a proto-galaxy (credit: NASA/JPL-Caltech/STScI/NAOJ/Subaru).

Table 1.1 Some galaxy types.

	Description	Intrinsic luminosity range (relative to Sun)	Mass range (in solar masses)	Diameter range (in thousands of light years)
Spiral	Flat disc, with spiral arms	0.1 billion to 100 billion	1 billion to 1 trillion	20–300
Elliptical	Smooth ellipsoidal shape	1 million to 1 trillion	1 million to 10 trillion	3–500
Irregular	No particular shape, lots of gas and dust	1 million to 1 billion	1 million to 100 billion	3–30
Starburst	Rich interstellar medium, lots of star formation	1 billion to 100 trillion	1 million to 10 billion	300–3000
ULIRG (Ultraluminous Infrared Galaxy)	Almost all emission in the infrared, very bright indeed	Larger than 1 trillion	1 trillion?	1000?

Table 1.2 Interstellar and intergalactic distances.

Mean Earth diameter	12.7 thousand kilometres
Mean Earth to Moon distance	0.384 million kilometres or 1.3 light seconds
Mean Earth to Sun distance	149.5 million kilometres or 8.3 light minutes; this distance is also known as an Astronomical Unit (or AU)
Mean Pluto to Sun distance	5.91 billion kilometres or 5.5 light hours
Alpha Centauri (the nearest star) to Sun distance	4.37 light years
Galactic Centre to Sun distance	27 thousand light years
Milky Way diameter	100 thousand light years
Milky Way to Andromeda galaxy distance	2.5 million light years
Milky Way to Virgo galaxy cluster distance	59 million light years
Visible Universe diameter	93 billion light years

measured in millions of light years, while the Universe as a whole is more conveniently measured in billions of light years, or similar units.

We'll be concerned with dust in galaxies, especially in the Milky Way which turns out to be a fairly typical galaxy. Some other galaxies have more gas and dust than the Milky Way, and others have less. The amount of dust compared to the amount of gas in the interstellar medium of galaxies is also an interesting quantity, and it may also vary from galaxy to galaxy. Some galaxies have higher dust-to-gas ratios than in the Milky Way, while others have lower ratios. Obviously, since some special stars are the source of dust in a galaxy (we'll discuss this in more detail in Chapter 4), galaxies that have had a major star-forming phase will have produced many of those special stars that create a lot of dust. Such galaxies will show high dust-to-gas ratios, while less active galaxies may be expected to have lower dust-to-gas ratios.

1.3   ACTIVITY IN THE INTERSTELLAR MEDIUM OF THE MILKY WAY

The interstellar medium of the Milky Way is an active place. On the one hand, gravity tends to pull matter together, creating clumps of dusty gas that in some situations may collapse and lead to the formation of new stars. On the other hand, stellar explosions (such as supernovae explosions that end the lives of massive stars) and stellar winds try to rip apart any structures in the gas. All this activity is taking place in the rotating disc of the galaxy, so all the structures within the galaxy tend to spin. This spinning also affects the kinds of structures that form in the interstellar medium. Another factor is the presence in the interstellar medium of a magnetic field. It acts on charged particles (ions and electrons) in the gas, and through collisions of ions with neutral atoms and molecules affects the bulk of the gas. The result of all these physical processes is that the interstellar medium is – as we said in the previous section – very clumpy and irregular, and the medium is highly turbulent on length scales from small to large.

The range in number densities, temperatures, and typical sizes of structures to be found in the interstellar medium is very great indeed. Large regions of interstellar space have such very low gas densities (meaning that gravity is weak) and very high temperatures (meaning the gas pressures are strong) that star and planet formation do not occur in them. For example, about half of interstellar volume is occupied by gas that has been shocked by supernovae winds, and is at enormous temperatures of about a million degrees. However, this gas has such a low density (about one thousandth of the average interstellar density) that the total mass of gas in these regions is (relatively) very small. Other regions are denser (similar to the average density of the interstellar medium) and cooler (but still hot, at about ten thousand degrees) and occupy much of the remaining volume of interstellar space. Almost all of the mass of the interstellar medium is found in the remaining few percent of the interstellar volume, in the form of cool clouds and clumps, with gas densities that are many times the average interstellar density. It is in these denser regions that interstellar chemistry and star and planet formation can occur, and where dust plays its important roles. In Table 1.3 we list some of these denser, cooler structures identified in the interstellar space of the Milky Way and other galaxies. Matter is cycled between these cooler structures, and between them and the hot, low-density structures.

Although dust is found in all of the interstellar regions, it is concentrated in those denser regions of interstellar gas, in particular the regions that in Table 1.3 are called *diffuse*, *translucent*, and *dark* clouds, and *star-forming regions*. These are the regions where dust has its most important roles and in the following chapters we shall describe those roles in detail.

Table 1.3 Some cooler, denser structures in the interstellar medium of the Milky Way.

Name	Number density[a]	Temperature[b]	State of hydrogen[c]
Diffuse cloud	100	100	Atomic and molecular
Translucent cloud	1000	30	Atomic and molecular
Dark cloud	10 000	10	Mainly molecular
Star-forming region	10 million	300	Mainly molecular

[a]The number density is the number of hydrogen atoms per cubic centimetre of space.
[b]The temperature given is the number of degrees Celsius above absolute zero (which is −273 °C, *i.e.*, 273 degrees below the temperature at which water freezes on Earth).
[c]Neutral hydrogen can be atomic, H, or molecular, H_2. Hydrogen can also be ionized, H^+, and a small amount of ionized hydrogen is present in diffuse and translucent clouds.

In very hot regions and in shocked gas, the dust is eroded by impacts of atomic nuclei and by collisions between dust grains, and so it may be at least partially removed in such regions.

Table 1.3 shows that diffuse, translucent and dark clouds are much denser than the average interstellar density. They are cool and mainly neutral. As we'll see in Chapters 6–8, they contain molecules that are crucial for the evolution of these regions, and dust plays important roles in establishing the chemistry that produces these molecules. In diffuse clouds, the most important molecules are molecular hydrogen (H_2) and carbon monoxide (CO). Dark clouds and star-forming regions are chemically rich; about two hundred different types of molecular species have been identified in these regions. These types range from fairly small molecules familiar from experiences in school laboratories, such as water (H_2O), formaldehyde (H_2CO), and ammonia (NH_3), to relatively complex organic molecules such as ethanol (C_2H_5OH), dimethyl ether (CH_3OCH_3), or glycolaldehyde (CH_2OHCHO). The simplest molecules have an essential role in star and planet formation, and dust is an essential partner in the processes that lead to the formation of these molecules. The larger molecules clearly have a potential connection with astrobiology. We'll discuss their link to dust in Chapters 7 and 8, where we show that the presence of dust is necessary for the formation of these complex organics in the interstellar gas. The huge variety of atomic and molecular species in the interstellar gas was discovered through spectroscopy; *i.e.*, the ability of atoms and molecules to absorb or emit (depending on density and temperature) radiation at particular wavelengths. You can find out a little more about spectroscopy in Box 1.2.

The purpose of this book is to show that dust is essential to make stars like the Sun, planets like the Earth, and for the formation of astrobiological molecules in interstellar space. We don't yet know the relevance of those molecules to life on planets. We'll describe these special roles of dust in the later chapters of this book. In Chapters 2 and 3 we'll discuss how we know that dust is present in the interstellar medium and how we can characterize it.

Box 1.2   Spectroscopy

Atoms, molecules, and solids can be considered to be made of relatively heavy positive nuclei together with much lighter negative electrons moving in orbits around them. The electromagnetic fields of radiation interact with the electrons; this means that the radiation may be absorbed or scattered. Atoms possess both *discrete* (individually distinct) and *continuous* absorptions. Discrete absorption removes radiation at a specific wavelength, and we can think of this as corresponding to electrons moving from one orbit to another one of higher energy. For example, gaseous sodium atoms can absorb in discrete lines at 589 and 590 nm and in some other lines too. If sodium atoms are irradiated by any radiation with wavelengths shorter than about 241 nm then the radiation is powerful enough to remove the electron entirely from its orbit around the nucleus. Since this happens for all radiation with wavelengths shorter than 241 nm, this causes a continuous absorption of radiation for wavelengths shorter than this limit.

Molecules show a similar behaviour, with both discrete and continuous absorptions, but there is a further complication. Molecules are complex structures composed of atoms bonded together, and these structures can vibrate and rotate. Vibration and rotation make the single discrete line of an atom break into a structure of many distinct discrete lines, so the line spectrum of a simple molecule such as carbon monoxide is very much more complicated than of an atom such as carbon or oxygen. The carbon and oxygen atoms are bonded together in a strong chemical bond to form a carbon monoxide molecule, and of course if the molecule didn't vibrate and didn't rotate then the molecule would have a spectrum of discrete lines corresponding to electronic transitions, just like an atom. But the bond is elastic, so that the carbon and oxygen atoms can vibrate, and the single line is replaced by a set of lines that correspond to increasing frequencies of vibration. Similarly, in a gas, the molecule is free to rotate, so each line corresponding to a particular vibration frequency is also split into many lines corresponding to increasing frequencies of rotation.

Molecules in a gas can make transitions between vibrational energy levels and also between rotational energy levels. For CO, transitions between the two lowest energy pure rotational energy levels correspond to radiation of wavelength 2.6 mm, while transitions between the two lowest pure vibrational levels correspond to radiation of wavelength 4.5 μm.

Solids generally show continuous absorption due to the large number of overlapping lines, but "absorption edges" also occur, corresponding to absorptions associated with the atoms in the solid. For example, when radiation at an appropriate wavelength irradiates a silicate containing magnesium, iron, silicon, and oxygen the absorption spectrum will show a broad underlying continuous absorption with superimposed narrow bumps and steps associated with individual atoms. In general, the response of any material to irradiation by an electromagnetic field is described by a quantity known as the *refractive index*. Refractive indices are usually measured in a laboratory experiment, and used in computations of the scattering and absorption of radiation by dust particles.

1.4   A SHORT HISTORY OF INTERSTELLAR DUST

The study of astronomy is as old as humankind, but the history of observations of interstellar dust is relatively brief. The first recorded remark that there might be something unexpected in the observations of rich fields of stars was made by William Herschel in 1784. He was making observations (by eye, but viewing through his large telescope) of regions of the sky that showed a very dense population of stars. He saw that within one such star-rich region there was a small zone in which there was apparently a complete absence of stars, and he famously noted (in German) "Here indeed is a hole in the Heavens!" He may have believed that there was a true absence of stars in that zone, but an alternative hypothesis is equally tenable: that the stars were not absent but that something between the stars and his telescope was obscuring those stars from view. We now know that this alternative hypothesis is correct. There is an interstellar 'fog', which can be thick enough to obscure the light of background stars, just as a severe fog in the Earth's atmosphere may obscure streetlights. This idea was confirmed in the 19th century by Edward Emerson Barnard who used photographic techniques to make images that were much more sensitive than those Herschel could make simply by looking through his telescope and sketching what he could see. Barnard could detect many stars not seen by Herschel and some of these were in the supposed "hole in the Heavens". Barnard's photographs showed that there were many structures in the interstellar medium that were capable of absorbing the light of stars.

This 'fog' was not only localized in well-defined regions but found to be generally distributed throughout interstellar space. In the 19th century, Wilhelm Struve counted the number of stars at different distances from the Sun, and found that these observations seemed to show that the number of stars per unit volume of space declined with distance from the Sun. This seemed unreasonable. Why should the Solar System be in this special position? Struve offered a more acceptable interpretation: that a *general interstellar extinction of starlight* was occurring, and this extinction made starlight weaker than expected, so that the estimated distances to the stars were too large and consequently the number of stars per unit volume too small. A similar conclusion was reached by Robert Trumpler in the 20th century from his observations of spherical associations of stars called *globular clusters*; the clusters seemed to be larger the further they were from the Sun; in fact, extinction makes them fainter and therefore apparently further away. Observations such as these indicated a general extinction of starlight throughout the interstellar medium. We now interpret the *local* extinction of starlight seen by Herschel (his "hole in the Heavens") and by Barnard as arising from the presence of a particularly dense cloud of gas and dust in the line of sight, which obscures the light of more distant stars.

Trumpler also showed that the amount of extinction towards a particular star varied with colour: observations made in blue light were more heavily extinguished than observations made in red light. Starlight is therefore "reddened" – *i.e.*, appears redder than it should – as it passes through the interstellar medium, and the amount of reddening is a measure of the amount of interstellar matter along the line of sight. It's very similar to the Earth's atmospheric reddening we observe when the Sun is setting, and the sunlight makes a long path through the Earth's atmosphere. The Sun's blue light is more strongly scattered than red light by matter along that path, and so we see the residual red light more strongly than the blue. In 1939, Joel Stebbins and colleagues showed that dust grains comparable in size to the wavelength of light were capable of causing this wavelength-dependent extinction. This conclusion strongly supported the idea that interstellar dust grains caused interstellar extinction, and implied that a range of grain sizes was required to account for the observations.

At the time of these discoveries, most astronomers wanted to study the stars themselves, and the interstellar extinction was simply an "irritating fog" that prevented them from doing their work more accurately. So, they devised a way of taking account of extinction by re-calibrating their observational data, allowing for the variation of extinction with colour (or, more accurately, with wavelength). Obviously, they needed to make a bigger correction to the blue end of the stellar spectrum than to the red end. Unfortunately for astronomers, the variation of extinction with colour (or wavelength) of starlight – the so-called *interstellar extinction curve* (ISEC) – wasn't a unique correction, but was found to depend on the line of sight, so the correction wasn't a simple adjustment for all lines of sight. Each line of sight had a different correction.

We show in Figure 1.4 a schematic diagram of a typical ISEC. In this diagram, the wavelength of the radiation is given on the *x*-axis, and the amount of extinction corresponding to the wavelength is shown on the *y*-axis. By convention, the wavelengths run from large values on the left to small values on the right, and the infrared, visible, and ultraviolet regions are indicated. The extinction is shown rising from small to large. The amount of extinction that is plotted in figures like this one is always normalized, so that all ISECs refer to the same amount of gas or dust. Differences between ISECs are then due solely to differences in the dust itself, and not to the amount of dust. Figure 1.4 indicates the fairly general behaviour that extinction is always small in the infrared, rises in the visible, has a "bump" in the near-ultraviolet, after which the extinction continues to rise. We'll have more to say about all these features in the following chapters, but it's a good idea to have a very general picture of the shape of an ISEC in mind as we proceed.

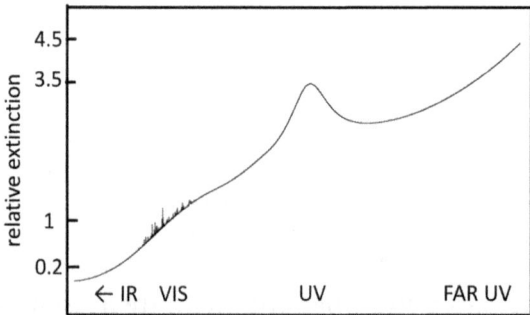

Figure 1.4 Characteristic ISEC diagram. This schematic diagram shows how the average extinction measured in the Milky Way galaxy varies with wavelength. At infrared (IR) wavelengths, the average extinction is small compared to that in the visible (VIS) region of the spectrum. Extinction increases in the visible region, and rises to a peak (often called the "bump") in the near-ultraviolet, then decreases before rising steeply into the far-ultraviolet (far UV), where it is typically three times as strong as in the visible.

It was soon realised that these ISECs were, in fact, a valuable data resource from which information about the dust causing the extinction could be obtained. We'll discuss ISECs as a source of information about the nature of interstellar dust in Chapter 3. We'll see there that ISECs in the optical – and ISECs extended into the infrared and ultraviolet regions of the spectrum – can be explained by the presence of an extinguishing agent in the form of small solid particles of a size comparable with the wavelength of the light being extinguished. So (as Trumpler had claimed) a range of particle sizes is required to account for extinction ranging from the infrared (wavelengths about a thousand nanometres or more) to the ultraviolet (wavelengths about a hundred nanometres or less). Thus, dust grains should have a comparable range of sizes.

One of the regions which Barnard discovered to be very heavily extinguished in the optical region of the spectrum (called B68 because it is the 68th member of his list), has also been observed at infrared wavelengths. The infrared radiation penetrates more easily than optical radiation (as indicated by Figure 1.4), so that one can confirm that stars really exist behind the opaque optical image. There really isn't a "hole in the sky", but simply a region of heavy extinction. We can see this very clearly in Figure 1.5.

There is one further piece of information that, historically, confirmed the idea that extinction is caused by small solid dust grains. In 1949, working independently, John Hall and Albert Hiltner discovered that starlight was often found to be polarized, and that the degree of polarization was generally greater for stars that showed a greater amount of reddening. The polarization evidently was linked to the extinction mechanism; the more dust along the line of sight to a star generally meant that there was also more polarization. A mechanism for polarization of starlight is differential extinction by partially aligned grains. Polaroid spectacles contain molecules oriented in such a way that they extinguish light with one plane of polarization more than another. Sunglasses are arranged so that they suppress the glare from reflected light, which is polarized. In a similar way, partially aligned asymmetric dust grains in the interstellar medium can act as the polaroid spectacles do and can polarize starlight by absorbing light in one plane of polarization more than another,

Figure 1.5   Barnard 68, (left) optical and (right) infrared images of a small
isolated dark cloud, viewed against the distant background of a
rich field of stars. In the optical image, the dark cloud is very dark,
and no background stars can be detected through the cloud. But
the infrared image shows that stars are in fact very abundant in
the background, and can readily be seen because the extinction
due to dust is much less in the infrared than in the optical region
of the spectrum (credit for both images: ESO).

so that the transmitted light is partially polarized. If you wish,
you can find a bit more information about polarization of starlight in Box 1.3.

By the middle of the 20th century the existence of interstellar
dust was well established. Since then, there has been a growing
recognition of the importance of this component of the interstellar medium. It can no longer be regarded merely as an irritating fog that obscures the light of background stars. Its roles in
interstellar chemistry, in star and planet formation, and in providing molecules that seem to provide a basis for astrobiology
are explored in the later chapters of this book. Interstellar dust
must now be regarded an essential component in galaxies like
the Milky Way.

1.5   SOME OTHER KINDS OF DUST THAT AFFECT OUR LIFE
ON EARTH

We'll claim in this book that dust in the Milky Way and other
galaxies is a very good thing, if stars and planets are to form
and life is to emerge. Of course, everyone is familiar with

Box 1.3   Polarization of starlight

What does it mean to say that light can be *polarized*? We know that light is a wave motion, so let's think first of a more familiar example of wave motion: gentle waves on water, like those you might get if you drop a pebble into a pond. Water waves move horizontally over the surface of the water, and the waves themselves are motions in the water in a vertical plane. These water waves can be said to be *polarized* in the vertical plane, since the motions of water molecules are always in the vertical plane. The waves move horizontally, while the water molecules move in a vertical plane.

Light waves are electromagnetic waves, and obviously they can move in any direction, not just horizontally. The waves occur in the electric and magnetic fields, and these fields are always perpendicular to each other and to the direction of motion of the light, see Figure 1.6. If light waves have their electric fields in a specific *fixed* plane, perpendicular to the direction of the light, as drawn in Figure 1.6, then the light is said to be *plane polarized* or *linearly polarized*. In general, a light beam contains a mixture of waves with electric fields in many different planes. A light beam with a mixture of waves covering equally all possible electric field directions is said to be *unpolarized*.

It's possible to start with unpolarized light and make it (at least partially) polarized. We can do that by extinguishing (or simply suppressing) light that happens to have electric vectors in one particular plane. Then the remaining light doesn't have electric fields covering all directions equally, and it is therefore partially polarized. This process is one of differential extinction, in which light with electric fields in one particular direction is preferentially removed. To see how this might be done, let's return to our simple water wave example.

Imagine that the water wave flows towards a fence of *vertical* sticks. The vertical gaps between the sticks don't prevent the water molecules from moving vertically, and so the wave can flow easily through the gaps between the sticks and continue past the fence. The wave may suffer a loss of some energy because of turbulence caused by the sticks, and the height of the

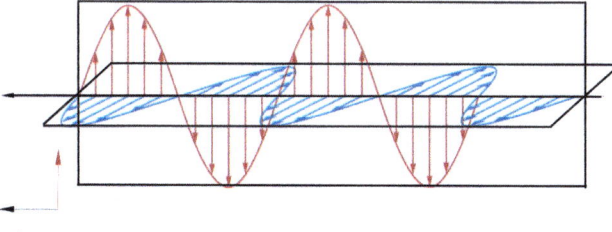

Figure 1.6   The schematic figure shows the variation of the electric (red line) and magnetic (blue line) fields associated with a light wave. The two fields are always perpendicular to each other and to the direction of travel, and the fields oscillate in phase with each other. If the electric field is always in the same plane, as shown, then the wave is said to be polarized.

(continued)

Box 1.3   *(continued)*

wave may therefore be reduced, but the wave is still present after passing through the fence. But if the fence is made of *horizontal* sticks, then as the water wave tries to penetrate the fence, the vertical motion of water molecules is severely restricted and so the water wave loses much of its energy. So, after passing through the fence, the wave is much weaker.

Polaroid is the common name given to plastic sheets containing crystals that are aligned in one particular direction. Light passing through the plastic sheet with electric fields along the alignment is more strongly absorbed than light perpendicular to the alignment. Thus, the light emerging from the sheet will be partially polarized.

In the interstellar medium, asymmetric dust grains of a size comparable to the wavelength of visible light can be partially aligned by interacting with the interstellar magnetic field. An interstellar cloud in which the dust grains are partially aligned is the equivalent of a polaroid plastic sheet, and starlight passing through the interstellar cloud can be weakly polarized. The observation of such polarization is regarded as strong evidence for the existence of interstellar dust grains of a size comparable to the wavelength of visible light, *i.e.*, with sizes of a hundred nanometres or more. These grains must be non-spherical.

---

various types of dust that occur on Earth and we know that these may have negative consequences. We have been made very aware of dust particles (or particulate matter) in our terrestrial environment and of the serious effects that they may have. *Particulates* in the atmosphere are called *aerosols*. Some of these are naturally occurring, such as sea spray, dust from volcanic activity, or ash from forest fires and grass fires. Others arise from human activities, which cause smogs and soots, together with chemical contaminants. Many of these *anthropogenic* dusts (*i.e.*, those created by human activity) have serious consequences on the health of human beings and on the climate of the planet. For example, diesel engines are known to produce "black soot" in the form of carbon particles, some of which have sizes less than 2.5 μm, together with many carcinogenic fairly small molecules including oxides of nitrogen, NO_x (implicated in respiratory conditions), and some larger polycyclic aromatic hydrocarbons. The small size of these sooty particles enables them to be drawn into the lungs and blood streams of humans, with a variety of very serious consequences for health. For example, it has been shown that death rates from lung cancer increase as the atmospheric abundances of these very small particles increase.

Aerosols also influence the climate on Earth by affecting the absorption of the Sun's radiation and the emission of radiation of longer wavelengths from the Earth. The effects are difficult to estimate, currently making the contributions to global warming/ heating from aerosols an uncertain factor in the estimates of climate change.

Particulates are also an important factor in the health of Earth's oceans. These particulates arise from the increasing use of plastics in many aspects of human life on Earth. Plastic debris finds its way into the oceans, which now contain not only large amounts of plastic bags and wrappings, but also microplastics with sizes that can range down to about ten nanometres. Microplastics and microfibres (shed from synthetic garments during washing) have now entered the food chain, and are found in many kinds of seafood with as yet unknown consequences.

By contrast with these concerns about various kinds of terrestrial dusts, particulates in interstellar space – observed from Earth – seem to be benign. We shall see in the following chapters that interstellar dust particles are of a similar size range and possibly contain similar compositions to many particulates found on Earth. Our most important finding will be that interstellar dust has important roles to play in catalyzing a rich interstellar chemistry, which influences the formation of stars and planets, and in creating pre-biotic molecules that may have a part to play in the creation of life itself.

1.6   DUST GRAINS AND COSMOLOGY

It's worth stating very briefly how dust should be placed in the general considerations of the origins and evolution of the Universe: *i.e.*, cosmology. When did dust appear in the Universe? Was the pre-dust Universe fundamentally different from the present Universe?

Observational evidence – such as the general expansion of all galaxies away from each other – supports the idea that our Universe began in a "Big Bang" some 13.8 billion years ago. The initial state in the Big Bang was extremely hot – almost beyond imagining. In fact, it was so hot that familiar types of matter – atoms, molecules, and (consequently) dust itself – could not exist. The Universe existed purely in the form of radiation. But this initial state wasn't stable: it had to expand. The expansion from that initial state caused the radiation to become less energetic, *i.e.*, to cool, and in this cooling phase the formation of some sub-atomic particles – such as protons, neutrons, and electrons – eventually occurred. As the cooling continued (though the radiation and these sub-atomic particles were still enormously energetic) atoms such as hydrogen (a proton and an electron bound together) and helium (two protons and two or three neutrons with two electrons bound together) were able to form from these sub-atomic particles. At first, these atoms didn't survive for very long because the powerful radiation tore them apart soon after they were formed, but as the expansion and the cooling continued, the ability of the radiation to destroy the atoms weakened and atoms began to survive and become abundant. Once this stage had been reached, massive clouds of hydrogen gas, with about 10% of helium atoms, began to accumulate, and within these primordial clouds the gas cooling became important and the effects of gravity caused some denser clumps to contract under their own weight. Ultimately the very first stars in the Universe were formed from these contracting clumps. At this stage, the matter in the Universe was almost entirely hydrogen and helium, and no dust existed. These first stars were formed *without* any possible influence of dust, but during their lives they converted hydrogen to carbon, oxygen, metals and other elements from which the first dust grains in the Universe eventually formed. It's thought that the first stars were formed when the Universe was about 180 million years old, or about 1.3% of its present age, so the earliest dust in the Universe must have appeared after that time. That means that for perhaps 98% of the time the Universe has existed, dust has been present.

Of course, we can't see those very first stars now. They exhausted their fuel, hydrogen, long ago. Theoretical studies suggest that these first stars were very massive indeed, had very short life times, and died in huge explosions similar to those we call *supernovae* – the death throes of massive stars in the modern Universe. In the early Universe, as in the present era, these stellar explosions populated the gas around them with the ashes of the "nuclear burning" that converts hydrogen to heavier elements, and with dust formed from those heavier elements. We'll discuss the origin of dust in the present era from various sources including supernovae, in Chapter 4.

So, dust has a long history in the Universe, probably just as long as the important elements heavier than hydrogen (such as oxygen, carbon, nitrogen, silicon) have existed. The abundance of dust in the Universe is steadily growing. In this book we shall describe the roles of dust in the formation of stars and planets and in the provision of molecules of biological interest, which might lead to life on planets or in space. In one sense, we might therefore come to regard dust in the Universe as "a machine for producing life".

CHAPTER 2

What Are Dust Grains Made of? How to Find Their *Chemical* Composition

It seems an impossible task! Just think about it: on average, over the whole of the disc of the Milky Way galaxy, dust grains of the size that cause interstellar extinction in visible light are on average about 100 m apart. They are typically a few hundred nanometres in diameter (a nanometre is a billionth of a metre, so the size of these grains is about one thousandth of the diameter of a human hair). Surely, it's impossible to say anything useful about the chemical composition of this highly dispersed and finely divided material? Don't despair! There are things we *can* do.

Firstly, we can try to keep track of the numbers of atoms available for incorporation into dust. Atoms that are in dust cannot be in the gas, and atoms in the gas cannot be in the dust. As we'll see, we can use this blindingly obvious statement to constrain the actual chemical composition of the dust. We could call this "interstellar book-keeping".

Secondly, although it's certainly a difficult task to try to collect some dust from interstellar space and bring it to Earth, there

Dust in Galaxies
By David A. Williams and Cesare Cecchi-Pestellini
© David A. Williams and Cesare Cecchi-Pestellini 2020
Published by the Royal Society of Chemistry, www.rsc.org

is one way in which we can actually get a reasonable amount of interstellar dust into our laboratories, where we can bring the whole apparatus of laboratory chemistry to bear on it. This *inter-stellar* dust is brought to Earth as inclusions inside *interplanetary particles*, that is, debris left over from the formation of the Sun and the Solar System. Fortunately, it's possible to distinguish between *interstellar* and *interplanetary* material.

Finally, the interstellar extinction curve (ISEC) has within its broad shape some much narrower absorption bands in the infrared part of the spectrum. These bands are created by the absorption of radiation by vibrating molecules or parts of molecules that are embedded in the dust grains. For example, strong absorptions that can be seen at 9.7 μm are caused by Si–O vibrations in silicates such as forsterite (which has the chemical formula Mg_2SiO_4) and fayalite (Fe_2SiO_4) or magnesium/iron mixtures of such silicates. In some interstellar regions there is absorption in a band centred at a wavelength of about 3 μm. This can appear in some dark interstellar regions as a relatively narrow band in the broad band extinction curve in the ISEC. It is caused when the radiation is absorbed by vibrations of the O–H bond in the water molecule (H_2O) embedded in water ice. We'll discuss more about interstellar ice in Chapter 8. There are also absorption bands at 4.7 μm that show in the ISEC of dark regions; these absorptions correspond to C–O vibrations in carbon monoxide trapped in the water ice. Obviously, absorptions like these tell us something quite specific about the chemical composition of the dust grains.

In Chapter 4, we'll look at dust formation in the envelopes of fairly cool stars and in the ejecta from supernovae. The models that astronomers have developed describe how dust is formed and predict its chemical nature, so these models also help us to understand the chemical nature of interstellar dust. However, in this chapter we want to look only at the *direct* evidence – such as it is – for the chemical nature of interstellar dust. In Chapter 3, using the conclusions of this chapter, we'll discuss how we can deduce its *physical* nature; that is: what is the distribution of sizes of interstellar dust grains? Are there many more small particles than large? Are grains made of a single material, with different grains being made of different materials, or are they composites, that is, bits of different materials stuck together?

2.1 INTERSTELLAR BOOK-KEEPING

Stars like the Sun are hot. Some stars are much hotter than the Sun and many are cooler. The temperature of the surface of the Sun is 5778 K, or 5778 degrees above absolute zero (see Box 2.1 for a discussion of absolute temperature).

This temperature is so high that solids such as dust grains cannot survive in the Sun. They would be very rapidly eroded by collisions of atoms and ions in the hot gas at that temperature. Therefore, dust grains don't exist in the Sun, and atoms that might be in dust grains in other circumstances are in the gas. Consequently, if we examine the spectrum of the Sun we should see emission lines in the spectrum that belong to all the elements that are in the Sun, and similarly for other hot stars. From the

---

Box 2.1 What is absolute temperature?

Everyone is familiar with the idea of temperature. We know that water on Earth is likely to freeze when the thermometer tells us that the temperature is zero Celsius, and (at sea level) that boiling water is at 100 Celsius (this commonly used unit of temperature is named after a Swedish astronomer of the 18th century). But what is temperature? What does it mean to say that one body has a higher temperature than another?

The fundamental idea of temperature is this: *temperature is a measure of the amount of energy contained in a body*. The body might be a gas, a liquid, or a solid. These are all made up of atoms and molecules, and if the temperature is high enough then collisions between atoms and molecules may release electrons. In a gas, atoms, molecules, and electrons are in constant motion; so they have energy of motion called *kinetic energy*. Molecules in the gas are also vibrating and rotating, so they have energy corresponding to vibration and rotation. At the other extreme, atoms and molecules in a solid are not free to move around, so they don't have kinetic energy, nor are they free to rotate because they may be too closely packed together, so they don't have rotational energy. However, they are free to vibrate, so they do have vibrational energy. So, the temperature of a solid measures the amount of vibrational energy in the solid.

Once we accept that temperature is equivalent to energy, so that hot bodies contain more energy than cool bodies then we can ask the question: what happens if we try to take all the energy away from a body? If the body has no energy, then we say it is at *absolute zero of temperature*, and write this as zero K, where K stands for a unit of temperature called a Kelvin, after the 19th century physicist Lord Kelvin. Obviously, it's impossible to have temperatures lower than zero K, because no more energy can be taken away from the body.

The unit K is defined to have the same size as the unit C. The temperature at which water freezes is 0 C on the Celsius scale or 273.15 K on the absolute scale. To change from absolute temperature to Celsius temperature all we need to do is subtract 273.15.

position of these lines we can determine the particular species of atom or ion to which they belong, and from the strength of the line (how broad and deep it is) we can say how many of those particular atoms or ions are present. So, the Sun can be used as a standard that tells us how many atoms or ions of any particular element there are, relative to hydrogen, the most abundant element. These numbers include all those atoms that in other situations – not in stars but in interstellar clouds – might be locked up in dust. Table 2.1 gives some of these elemental abundances relative to one million H atoms.

Hydrogen is by far the most abundant element, and so it is convenient to show in Table 2.1 the abundances of elements relative to hydrogen. The only element that has a relative abundance close to that of hydrogen is helium (curiously, helium was first discovered by a line in the solar spectrum that was detected during a solar eclipse, before the element was known to exist; the name comes from the Greek word for the Sun: *helios*). There are roughly ten hydrogen atoms for each helium atom in the Sun and in all stars; however, helium doesn't make molecules very easily. The chemically important elements oxygen, nitrogen and carbon have much lower relative abundances. They are thousands of times less abundant than hydrogen; other elements have even lower abundances. We can take the solar abundances shown in Table 2.1 as a standard for the Milky Way galaxy. Other galaxies have an equivalent table to that of Table 2.1, but those numbers are not hugely different from those for the Milky Way.

Let's now keep track of where the atoms are in the interstellar medium. We'll restrict our discussion to diffuse clouds, as we defined them in Chapter 1. These low-density, cool clouds are chemically less complicated than denser, colder clouds, and it's easier to keep track of where the various atoms are. For example, the element carbon in the gas of a diffuse cloud might exist as an atom, C, or as its ion, C^+ (where one electron has been removed from the atoms) or combined in a molecule, such as carbon monoxide, CO, or even methanol, CH_3OH (carbon can makes lots of varieties of molecules, as we know from biochemistry on Earth, and it also makes a wide variety of carbon-based molecules in the interstellar medium, as we'll see in Chapter 6). In fact, there aren't many carbon-containing molecules in diffuse clouds, and nearly all the carbon is in carbon ions. If some of the carbon in the form of carbon ions seems to be missing from the gas of the diffuse cloud, then our basic assumption is that these missing atoms must be in the dust grains. The same idea applies to all other elements.

So, what we have to do is use the background star that indicates the presence of extinction in the diffuse cloud, and from the absorption lines in its spectrum find the abundances of the various elements in that diffuse cloud. What we find is quite surprising: almost all elemental abundances relative to hydrogen for diffuse clouds are found to be less – in some case much less – than in the standard solar abundances shown in Table 2.1. This is found to be the case by studying abundances in various diffuse clouds, and by looking towards many different background stars. There are fairly small differences between different lines of sight, but the general result is clear: elemental abundances in diffuse clouds are generally less – and in some cases very much less – than in the standard solar abundances. Where have all those missing atoms gone? Our assumption is that the missing atoms are now in the dust grains. We show in Figure 2.1, a bar chart for the average diffuse clouds in the interstellar medium of the Milky Way, giving the fraction of atoms remaining in the interstellar gas. For example, for carbon, roughly half of the carbon atoms remain in the gas, while for oxygen about two-thirds remain in the gas. Almost all the nitrogen is in the gas.

On the other hand, most of the magnesium atoms are 'missing'; only about 4% of them remain in the gas, so about 96% of them are absent from the gas. It's a similar story with silicon and iron. The fractions of these elements remaining in the gas are only 7% and 1%, respectively. Where are the missing atoms? We assume that they are in the dust grains.

The information in Figure 2.1 is perhaps slightly misleading in this form. Although about half the carbon atoms remain in the gas, while nearly all the magnesium is missing, there is much more carbon available to begin with, as we can see in the table of standard solar abundances, Table 2.1. A better way to present this information, from the point of view of the composition of the dust grains, is to concentrate on the missing atoms. The abundances of the missing atoms are presented in Table 2.2, for some elements of possible interest.

Table 2.1 Relative abundances of some chemically important elements in the Sun, relative to hydrogen.

Element	Hydrogen	Carbon	Nitrogen	Oxygen	Magnesium	Silicon	Iron
Relative abundance	1 000 000	214	62	575	36	32	33

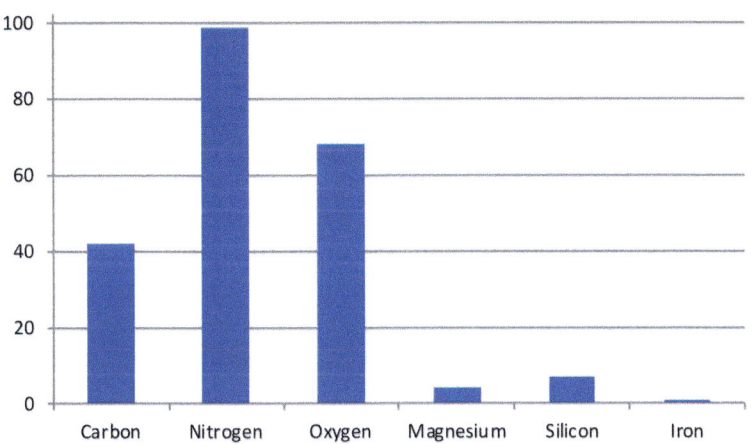

Figure 2.1 A bar chart showing the percentage of various elements remaining in the interstellar gas in diffuse clouds of the Milky Way.

Table 2.2 Relative abundances of some elements available for making interstellar dust.

Element	Hydrogen	Carbon	Nitrogen	Oxygen	Magnesium	Silicon	Iron
Relative abundance of missing atoms	1 000 000	123	0	186	35	29	33

Table 2.2 enables us to make an informed guess at the chemical nature of interstellar dust grains. The high abundance of oxygen relative to silicon suggests that silicates, that is, molecules containing the SiO_4 radical, could be abundant in the form of magnesium and iron silicates, Mg_2SiO_4 and Fe_2SiO_4, or some composite containing both magnesium and iron. We can see from Table 2.1 that there would still be some oxygen left over, and the excess oxygen could be taken up by metal oxides such

as FeO, MgO, or other oxides, if enough magnesium and iron are present. Some of the excess oxygen may, in some circumstances, be in the form of water, H_2O, but this water is less strongly bound together than species such as silicates.

While carbon can combine with many species, the large amount of carbon available for dust suggests that carbon particles may be present. Sooty particles are known to form in many situations including combustion, and are mainly solid carbon. Carbon combines readily with hydrogen, and given that there's lots of hydrogen present, it seems likely that some of the carbon will be present as hydrocarbon molecules. There is also evidence that at least some of the carbon is in a form that may be similar to graphite.

Of course, a great deal of very careful observational work is needed to support these general conclusions, but it is surprising (and very satisfactory) that the simple idea of "counting atoms" in the Sun and in diffuse clouds can lead to rather specific ideas about the chemical composition of interstellar dust in diffuse clouds.

In summary, our considerations suggest that interstellar dust grains in diffuse clouds in space are probably composed of carbons, hydrocarbons, silicates, and oxides. As we'll see, other approaches come to similar conclusions.

2.2 INTERSTELLAR DUST GRAINS ON EARTH?

Wouldn't it be helpful if we could find samples of interstellar dust on Earth? Then, by examining these samples in suitable laboratories we could determine not only the chemical composition but also the physical structure of that dust. Problem solved! The basic answer to the question is (rather surprisingly) "Yes, this can be done, but it is complicated!"

We know that interstellar gas and dust flow into the Solar System, more or less in the plane defined by the orbits of the planets around the Sun. Some decades ago a spacecraft, *Ulysses*, flew high over the poles of the Sun, far away from the debris left over from planet formation, and detected a flux of particles. So, there is no doubt that true interstellar dust is entering the Solar System. *Ulysses* was capable of determining the masses (and therefore approximate sizes) of the particles, and found that the dust it could detect was generally much larger than the dust grains that cause interstellar extinction. The ISEC computed by cosmic dust expert Bruce Draine for the grains detected by *Ulysses* is quite unlike the ISEC for the average Milky Way extinction. Evidently, the *Ulysses* large grains aren't responsible for the average interstellar extinction.

However, there are probably several reasons why the smaller grains weren't detected. For example, while all grains would be affected to some extent by the Sun's magnetic field, it is the smaller, less massive, grains (like those that cause interstellar extinction) that would be more strongly deviated out of the flow.

Another space mission, *STARDUST*, flew in 2004 through the coma (the dust tail) of a comet called Comet Wild 2 and collected thousands of particles from that object. This mission was very cleverly designed to return its collected samples to Earth in 2006, so the collected particles could be examined in the laboratory. Nearly all of these were large particles, very much larger than those particles (that is, comparable in size to the wavelength of visible light) needed to supply interstellar extinction, and these collected comet particles were crystalline, that is, they had a regular structure. The fact that they were crystalline seemed to indicate that these grains had at one time been hot (unlike our expectation for interstellar dust). Heating amorphous materials (those that have a random structure) can allow them to adopt a crystalline form (the process of conversion from amorphous to crystalline is called *annealing*, and is brought about by heating). So nearly all these cometary particles collected by the *STARDUST* mission are regarded as dust particles that have been formed during the energetic processes occurring in the creation of the Solar System. Out of many thousands of particles collected by the *STARDUST* mission, only seven were thought to be from interstellar space. Three of them seem to be large particles similar to dust particles detected by *Ulysses*, while the other three are much smaller.

The main types of dust grains detected by *STARDUST* in the tail of Comet Wild 2 include graphite, silicates, oxides, and silicon carbide. There are also even a few very tiny diamond grains, each containing just a few thousand carbon atoms (far too small to see, unfortunately, so it's no good as interstellar jewellery – one can't impress one's friends with these diamonds).

A few years ago, an instrument called the Cosmic Dust Analyzer on board a spacecraft called *Cassini* revealed some interstellar dust grains close to Saturn, coming from our immediate interstellar neighbourhood – the so-called *local interstellar cloud*. Thirty-six of these grains were collected and analyzed, and their mass distribution and elemental composition were determined. The results were in agreement with those of the *STARDUST* mission – with one important exception. While the *STARDUST* samples contain sulfides, there was no indication of sulfur in the *Cassini* dust grains. When compared with the elemental composition of interstellar dust grains preserved in 4.6 billion year old meteorites, these observations indicate that the major population of interstellar dust grains is different from pristine circumstellar stardust. Evidently, the composition of some dust grains may change during the planet-forming process.

So, while interstellar grains are entering the Solar System now, it isn't easy to detect these new arrivals, and evidently even if we can detect them we don't get a complete sample. Is there an alternative to searching for isolated interstellar grains within the Solar System? Fortunately, there is.

Perhaps it's better to recognise that the formation of the Solar System has been an energetic process that turned interstellar dust and gas into the Sun and planets, comets, asteroids, and interplanetary dust grains. (We'll discuss these topics in more detail in Chapter 9.) Is it possible that some true interstellar grains could have survived relatively unchanged during the formation of the Solar System and be trapped in solid material, especially in interplanetary dust grains? Fortunately, this trapping of interstellar grains within larger interplanetary grains does occur. For convenience, Box 2.2 summarizes the various types of solid material in the Solar System; our main interest here, of course, is in the dust component. The formation of stars and planets and the roles of dust in those processes will be discussed in more detail in Chapter 9. In Box 2.2 we simply describe the forms of the solids found in the Solar System.

To search for interstellar dust grains within interplanetary matter, then the important question is this: can we distinguish between interplanetary dust material and true interstellar

Box 2.2   Solid material in planetary systems

Detections have now been made of about four thousand planets in orbit around stars in the Milky Way galaxy outside the Solar System, just as planet Earth and all the other planets in our planetary system orbit the Sun. So far, we don't know a great deal about planetary systems other than the Solar System. Observations seem to suggest that the Solar System isn't very typical of all those external systems that have been observed. However, we do know quite a lot about the Solar System so in this box we'll mainly describe the various forms of the solid material that we find in our own backyard, and which we expect to be replicated in other planetary systems.

Planets

Most of the mass in the Solar System outside the Sun is, of course, in the planets. Mercury, Venus, Earth, and Mars are closest to the Sun and are called the *terrestrial* planets, while Jupiter and Saturn are located further from the Sun and are the *gas giant* planets. The planets most distant from the Sun, Uranus, and Neptune, are called the *ice giant* planets. In the process of star and planet formation, the interstellar medium is transformed from its normal low-density state in interstellar diffuse clouds into dense planetary material. Planets close to the Sun are sufficiently warm that much of the interstellar gas is lost, while planets further away are cool enough to retain these gases as liquids or ices. Images of Mercury, Jupiter, and Uranus illustrate these planetary stages; see Figure 2.2.

It is already clear from observations that external planetary systems also include both massive planets like Jupiter and less massive planets like the terrestrial planets. Many super-Earths are detected, with masses up to 10 Earth masses.

Asteroids

The region of the solar system between the orbits of Mars and Jupiter is the *asteroid belt*, in which solid bodies – *asteroids* – ranging in size from about one metre to about one thousand metres are found. The larger asteroids are regarded as *minor planets*. There are literally millions of asteroids in this region of the Solar System. Asteroids are the debris left over from the formation of planets. They are generally carbon-rich, or iron, or silicate materials. Asteroids are all in orbit around the Sun, just like the planets, but they do occasionally collide with each other and with planets. Planet Earth has been impacted by asteroids many times. In a recent example, the so-called *Tunguska event* on 30 June, 1908 flattened 2000 km^2 of Siberian forest in an explosion in Earth's atmosphere caused by an asteroid's close approach to Earth. The explosion is reckoned to be equivalent to about ten megatonnes of TNT (similar to a large hydrogen bomb). The impacts of millions of asteroids in the asteroid belt on each other, continually grind asteroids into smaller and smaller pieces and are the main source of interplanetary dust in the Solar System. Asteroids are generally irregularly shaped and heavily cratered; see Figure 2.3.

(continued)

Box 2.2   *(continued)*

Since the interstellar gas is well mixed, the interstellar cloud from which the Sun and planets formed incorporates interstellar gas and dust from many different sources (we'll discuss these sources in more detail in Chapter 4). Hence, planets and asteroids formed from this gas are well-mixed material and the interplanetary particles arising from the destruction of asteroids in collisions don't represent any particular source of dust.

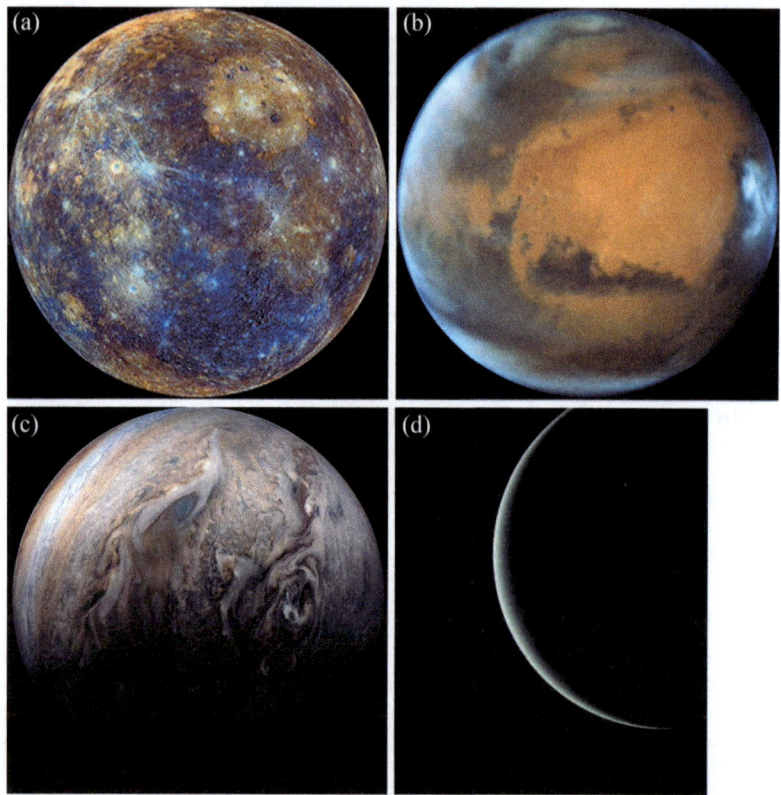

Figure 2.2   Planetary images. (a) Mercury (credit: NASA/Johns Hopkins University Applied Physics Laboratory/Carnegie Institution of Washington). (b) Hubble Space Telescope photo of Mars taken when the planet was 50 million miles from Earth on May 12, 2016 [credits: NASA, ESA, the Hubble Heritage Team (STScI/ AURA), and Z. Levay (STScI); Acknowledgement: J. Bell (ASU), and M. Wolff (Space Science Institute)]. (c) Jupiter captured by NASA's *Juno* spacecraft as it performed a close pass of the gas giant planet. Some bright-white clouds can be seen popping up to high altitudes on the right side of Jupiter's disc (credit: NASA/JPL-Caltech/SwRI/MSSS/Kevin M. Gill). (d) Image of Uranus taken by Voyager 2 on January 25, 1986 reveals its icy blue atmosphere, which it turns out is both deadly and smelly (credit: NASA/JPL).

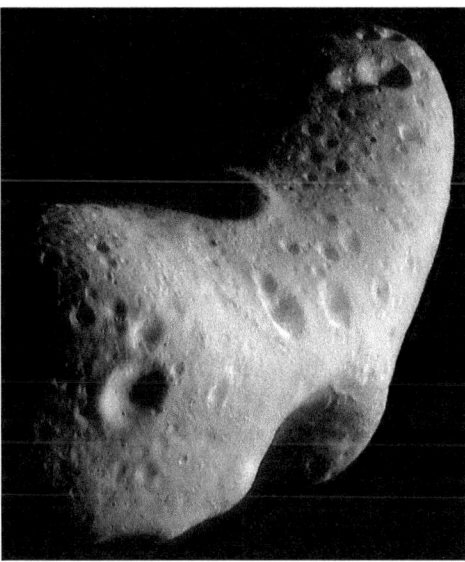

Figure 2.3   Image of an asteroid: Mosaic of Eros' Northern Hemisphere. The mission NEAR-Shoemaker succeeded in closing in with the asteroid and orbited it several times, finally terminating by touching down on the asteroid on February 12, 2001. [Credit: NASA/JPL/JHUAPL, The Near Earth Asteroid Rendezvous – Shoemaker (NEAR-Shoemaker).]

Comets

Comets have nuclei of up to about 10 km in size composed of dust, ice, and rocks. When they approach the Sun they suffer outgassing, which leaves tails of gas and dust that can be dramatic visual objects; see Figure 2.4.

Long period comets originate in the Öpik–Oort Cloud, a huge spherical cloud of gas and dust centred on the Sun and extending halfway to the nearest stars, several light years from the Sun. Long period comets probably have periods of hundreds or thousands of years. *Short period* comets originate in the *Edgeworth–Kuiper* belt, a circumsolar disc extending from the orbit of Neptune outwards. They generally re-appear within about one hundred years or so. Comets contribute dust to interplanetary space. Some of this dust may be interstellar dust; that is, some of the dust emerging from comets retains a "memory" of the place in which it was formed.

Meteoroids and Interplanetary Dust

Small objects in space, less than a metre in size are called *meteoroids*. When these enter the Earth's atmosphere at high speed, the friction causes them to burn up and at night we may see a short-lived streak of light known as a *meteor*. If part of a meteoroid survives the journey through the atmosphere and arrives at the Earth's surface, then the solid remnant is called a *meteorite*. It is estimated that about 15 000 tonnes of meteorites arrive on Earth each year. Meteorites are an extremely important source of information about the complex chemistry that may occur on dust in interstellar space.

(*continued*)

Box 2.2 *(continued)*

We'll discuss this in more detail in Chapter 10. Many of the collected meteorites were found in ice sheets in Antarctica. Confusion with and contamination by terrestrial material is minimized in these environments.

The distribution of solids continues towards much smaller sizes, and these particles are called *micrometeoroids* or interplanetary dust grains. Evidently, the Solar System is rich in solids of a wide range of sizes, and it's now clear that star- and planet-forming regions elsewhere in the Milky Way galaxy are rich in debris left over from the formation process. Figure 2.5 shows a false-colour image of the region around a young star called HL Tauri, of emission in the infrared from this debris. The material appears to be constrained into rings, presumably by newly-formed planets.

The Solar System probably passed through a phase similar to what we see in the HL Tauri image.

Visitors from Other Solar Systems

A visitor from interstellar space passed through the Solar System recently, and was detected on October 19, 2017 by Robert Weryk, 40 days after its closest approach to and heading away from the Sun. It did not show a cometary tail. It is moving so fast that its velocity exceeds the escape velocity of the Sun, so it cannot be a member of the Solar System and must therefore be from interstellar space. It is an elongated body, and is tumbling. There must be many such visitors. This visitor has been given the delightful name of Oumuamua, or – more prosaically – 1I/2017 U1.

Figure 2.4   This close-up view of comet Hartley 2 was taken by NASA's EPOXI mission during its flyby of the comet. It was captured by the spacecraft's Medium-Resolution instrument (credit: NASA/JPL-CalTech/UMD).

Figure 2.5 Proto-planetary disc around HL Tauri [credit: ALMA (NRAO/ ESO/NAOJ); C. BROGAN, B. SAXTON (NRAO/AUI/NSF)].

grains? Sometimes, interstellar dust grains embedded in other Solar System materials may be chemically distinct from the interplanetary dust material, so it's clear that these are separate objects formed of different materials. If so, then these interstellar dust grains can be separated from the embedding interplanetary material and examined. But such a clear distinction between interstellar and interplanetary materials is not often the case.

Fortunately, there's a reliable alternative method for distinguishing between true interstellar dust grains and the interplanetary dust material in which they are embedded. This is done by studying the isotopes in the material. See Box 2.3 for a discussion of elemental isotopes.

Interstellar grains are formed (as we'll discuss in more detail in Chapter 4) in outflows from some cool stars and in the explosions of supernovae and novae. Each type of source has a characteristic signature in terms of elemental isotopes. For example, carbon atoms occur naturally in isotopes that are chemically

Box 2.3   Elemental isotopes

Elements occur naturally as *isotopes*: these are atoms with the same chemical properties but different masses. For example, hydrogen atoms occur with masses one, two, and three atomic mass units. We write these as ^{1}H, ^{2}H, and ^{3}H; they are called *protium, deuterium*, and *tritium*, respectively. The simplest of these three isotopes is ^{1}H, protium, which can be considered as being made up of one proton (which has one positive charge) as a nucleus, with one electron (which has one negative charge) in orbit around the proton. Almost all the mass is in the proton. In deuterium, the nucleus contains not only one proton but also one neutron. The neutron has no charge, but does have a mass that is very close to that of the proton. So, the mass of a deuterium atom is two atomic mass units. Similarly, a tritium atom has one proton with two neutrons in its nucleus, so the mass of a tritium atom is three atomic units.

No other hydrogen isotopes exist, because the addition of neutrons to the proton begins to make the atom unstable. In fact, even tritium is unstable and decays so quickly that half of the number of ^{3}H atoms that you start with will have decayed in 12.3 years. This is called the *half-life* of tritium. However, protium and deuterium are stable; they don't decay. Protium is about 10 000 times more abundant than deuterium, and tritium is even less abundant than deuterium.

Carbon occurs with stable isotopes ^{12}C (which has six protons and six neutrons in the nucleus, and six electrons), and ^{13}C (which has six protons, seven neutrons, and six electrons). On Earth, ^{13}C is present with an abundance of about one percent of ^{12}C. ^{14}C has six protons and eight neutrons, with six electrons, and is radioactive with a half-life of 5740 years. The decay of ^{14}C in any carbonaceous object enables an age for the object to be estimated. This is called *radiocarbon dating*.

All elements arise in various isotopic varieties. For example, nitrogen has two stable isotopes, ^{14}N and ^{15}N, each N-atom containing seven protons and seven electrons, together with seven or eight neutrons, respectively. On Earth, ^{15}N has an abundance of about one third of one percent of ^{14}N. There are also some radioactive isotopes of nitrogen. Oxygen has three stable isotopes ^{16}O, ^{17}O and ^{18}O, each O-atom containing eight protons and eight electrons, together with eight, nine, and ten neutrons, respectively. On Earth, ^{18}O has an abundance of 0.2 percent of ^{16}O, and ^{17}O is even less abundant. There are a number of radioactive isotopes of oxygen. Many elements, like those mentioned above, are present with just one isotope dominating the abundances. However, some elements have isotopes present in comparable abundances (*e.g.*, chlorine isotopes ^{35}Cl are just three times more abundant than ^{37}Cl isotopes, while bromine isotopes ^{79}Br and ^{81}Br are almost equal in abundance).

The importance of isotopes in astronomy is that the locations in which elements are formed (various types of stars and explosions) through nuclear chemistry are different, and have different densities and temperatures. The products of the nuclear chemistry depend on those conditions, and so the relative abundances of the isotopes of a particular element are signatures of the location. Astronomers can determine the relative abundances of isotopes for one particular source, say, a supernova. Dust grains formed from supernova ejecta will carry that information in their atoms. It tells astronomers in a precise and specific way the origin of those grains. For example, carbon grains from supernovae have $^{12}C/^{13}C$ ratios ranging from about 10 to 10 000, while carbon grains from the envelopes of cool stars have $^{12}C/^{13}C$ values generally around 10. The supernovae grains have $^{14}N/^{15}N$ ratios ranging from about 10 to 100, while the cool star grains have $^{14}N/^{15}N$ values in the range of about 100 to 10 000. Using information like this, astronomers can determine the origin of particular grains and find out if they are truly interstellar dust grains.

identical but have different masses, 12, 13 (which are stable) and 14 (which is radioactive and decays). On Earth, ^{13}C is only about 1% of carbon. But the isotopic ratio of $^{12}C/^{13}C$ observed in specific astronomical objects may vary by a factor of about ten thousand. So, rather surprisingly, by examining the ^{12}C and ^{13}C abundances in a sample of material, it's possible to distinguish between grains of a particular type (for example, carbon) that originate in a nova, for example, from those that originate in a cool star envelope.

Of course, since the interstellar medium as a whole is well mixed, the cloud of gas and dust from which the Solar System was formed contained a mixture of grains from a variety of sources, and these were then processed into planets, comets, asteroids and interplanetary dust particles, and their isotopic ratios have an average value. But individual interstellar grains that have somehow survived, retain the isotopic signatures of their individual origin. In this Solar System context, the interstellar grains that have survived the star- and planet-formation processes unscathed are sometimes called *stardust*. It's a pleasing name, and since this dust contains the elements from which Earth and all it contains are made, it reminds us of our connection to the stars of the Milky Way. We are truly part of the Milky Way galaxy, not detached observers.

The conclusion from the *STARDUST* mission and from studies of interstellar and interplanetary material in meteorites supports the conclusions made from "interstellar accounting" (Section 2.1). Interstellar grains include silicates and carbons. The studies of meteorites show that this material is also very rich in organic molecules, from quite simple hydrides such as methane (CH_4) and acetylene (C_2H_2) to other molecules familiar from school chemistry, such as methanol (CH_3OH) and formaldehyde (H_2CO), and continuing on to much larger, more complex molecules. We'll meet these larger species in Chapter 8, and also in Chapter 10 where we'll discuss the relation of such molecules to astrobiology.

2.3   INFRARED SPECTROSCOPY OF REMOTE INTERSTELLAR DUST

We described the general shape of ISECs in Chapter 1, for lines of sight through cold diffuse interstellar gas clouds. As we have seen in Figure 1.4, in broad terms these curves show that extinction is rather low in the infrared region, rises steadily though the visible region from red to violet, and continues to rise into the ultraviolet, often with a distinct and very broad "bump" peaking at a wavelength of 217.5 nm. We have also noted that these ISECs are all broadly similar. They all show these characteristics, but they differ in details from one line of sight to another.

However, if the ISECs are examined carefully, then some additional features can be seen in the infrared part of the spectrum. These features are narrow compared to the broad structure of an ISEC, and certainly narrow compared to the width of the "bump". The strengths of these additional features may vary from one line of sight to another, but their positions in wavelength remain the same in all ISECs. That's an important fact. It tells us that these relatively narrow features are associated with the actual materials of which the dust grains are composed. Some examples of infrared spectra along lines of sight to different stars are shown in Figure 2.6. These spectra, which have been taken through rather dense interstellar clouds, show strong absorptions near 3 μm and 10 μm. They also show some other weaker features, but not necessarily in common.

These observations suggest one more method of determining the chemical composition of dust grains. We can make measurements in the laboratory of the infrared spectra of materials that are possible candidate materials of which interstellar dust is composed. Then we can compare the position and shape of the laboratory features with what we see in the astronomical spectra. If there is a match between the two, then we have a strong indication that the laboratory material is closely similar to a component of the dust grain material.

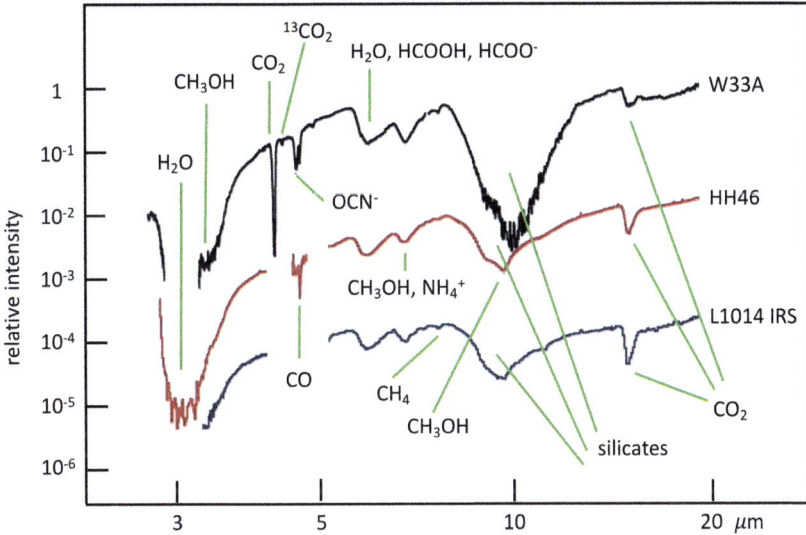

Figure 2.6 High-resolution infrared spectra towards several stars in the Milky Way galaxy, along lines of sight through dense cold clouds (data taken from K. I. Öberg *et al.*, *Astrophysical Journal*, 2011, **740**, 109).

This description suggests that it's a simple procedure, but in fact it is rather complicated because solids come in many forms, even if they have similar chemistries. These forms are affected by the way in which the materials are made. For example, the temperature of formation of the solid and the speed of cooling from a high formation temperature can affect the actual structures quite significantly.

The strong absorption at a wavelength of about 10 μm appears in all ISECs [remember that the visual spectrum occupies wavelengths between about 0.4 (violet) and 0.7 (red) μm, so this 10 μm absorption feature is well into the infrared region]. Most silicates show absorption at wavelengths near 10 μm, so it seems likely that silicates – which seem to be probable candidates of one component of dust materials, as discussed in Sections 2.1 and 2.2 – will be responsible for the absorption at 10 μm. The absorption is attributed to the excitation by the absorbed radiation energy of vibrations in the Si–O bond that exists in all silicates, such as Mg_2SiO_4 or Fe_2SiO_4, or in silicates with a mixed magnesium/iron composition. While all these silicates have absorption near 10 μm, the detailed shape and position depends on the actual chemical composition and on the physical nature of the silicate, such as whether it is amorphous or crystalline. Without going into details of the comparison between the astronomical and laboratory spectra, we can certainly infer that some silicate material is present in interstellar dust. This is entirely consistent with our deductions from elemental abundances and from studies of interstellar dust grains found embedded in interplanetary dust.

The absorption feature near 3 μm is attributed to excitation of the O–H bond in H_2O bound on dust grains in the form of water ice. Evidently, there can be a lot of water ice associated with interstellar dust. However, the ice doesn't appear in diffuse clouds, only on dust grains in the denser, darker regions of interstellar space. We can think of dust grains in diffuse clouds as being 'bare' silicates and carbons, so that atoms and molecules from the gas that collide with the grains actually hit a surface of (probably) a carbon or a silicate. Evidently, icy coatings accumulate on dust grains in very dark regions of space. These icy mantles turn out to be very important for the chemistry. We'll discuss the properties and chemistry of ice-coated dust grains in Chapter 8.

So far, we have concentrated on absorption by very cold dust. But dust sometimes becomes warm, for example, when the dust grains are close to a hot star, but not so close that they are destroyed. When the dust grains sufficiently warm, say as warm as the reader of this book, the grains become *emitters* of radiation rather than absorbers. Very small grains, which we can consider to be large molecules, are heated more easily than larger grains, so the emissions from these small grains/large molecules (containing, say, a few hundred atoms) can be prominent near to bright stars. The main features are at 3.3, 6.2, 7.7, and 11.3 μm. These features correspond reasonably well with those from molecules called *polycyclic aromatic hydrocarbons*, or PAHs. These molecules consist of planar pieces of a graphite sheet, in which the carbon atoms are bonded in hexagonal rings, with hydrogen atoms around the edges. Such structures are called *graphenes*. The structures of some examples of PAHs are shown in Figure 2.7.

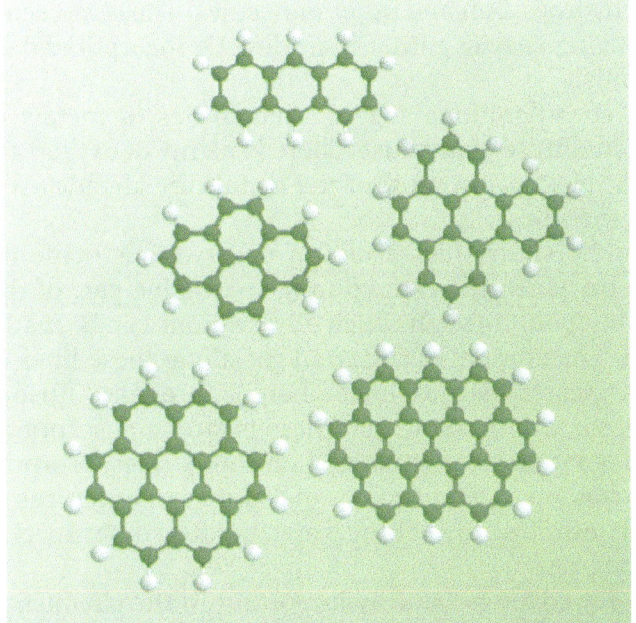

Figure 2.7   Structures of some examples of PAH molecules.

These emissions indicate that in addition to other dust materials, PAHs are also present in the interstellar medium.

2.4   WHAT DO THESE METHODS SUGGEST AS THE CHEMICAL NATURE OF INTERSTELLAR DUST GRAINS?

All the methods that we have described seem to agree quite well. Silicates of some form are strongly indicated as an important component of interstellar dust in diffuse clouds. But silicates form a rather varied class of materials, and the information so far obtained isn't specific about the precise nature of the material. Indeed, perhaps silicates change their structures in the interstellar medium, compared the structures we know from laboratory work.

There seems no doubt that carbon is an important component of interstellar dust in diffuse clouds. This may be partly like graphene (flat sheets of carbon arranged in hexagonal rings), in part an amorphous sooty material, and in part molecular in the form of hydrocarbon and other molecules. These molecules may be free-flying species rather than directly incorporated into the dust grains.

Some considerations suggest that oxides of metals may be present in diffuse cloud dust. There is plenty of oxygen available in diffuse clouds, but most of the metals are already used up in making the silicates.

Figure 2.6 shows the results of observations made on lines of sight on which the extinction in the visible part of the spectrum is high but through which observations can be made in the infrared. The amount of dust and gas along these lines of sight is much greater than on lines of sight through diffuse clouds. These observations show that there is strong absorption in the infrared near wavelengths of 3 μm, and this absorption has been attributed to the presence of coatings of water ices on dust grains. These ices are not pure water, because other species seem also to be present.

These icy coatings are very important in the chemistry of the interstellar medium, and will be discussed in more detail in Chapter 8. All we need do now is note that the two regions of space are different. In diffuse clouds we seem to have bare grains, so any atoms and molecules striking the grains will encounter the silicate or carbon surfaces that we have indicated in this chapter. On the other hand, denser darker regions lead to the formation of icy molecular mantles on grain surfaces. Reactions on and in these ices are important, as we shall see.

CHAPTER 3

What Is the Structure of Interstellar Dust Grains? How to Find Their *Physical* Composition

If we want to understand the contribution of interstellar dust grains to major events in the interstellar medium – and obviously that's our main aim in this book – then we need to know not only what the dust grains are made of (and we have made some progress in that regard in Chapter 2) but also how these materials are distributed in dust grains. For example, how many dust grains are there in total? What is the size range of these dust grains? How many small dust grains are there compared to large dust grains? Are the dust grains composites of several materials, or is each dust grain formed of a single material? It turns out that the best way to find answers to these questions is by using the interstellar extinction curve, the ISEC, as a source of information.

These questions significantly affect how dust grains influence their environment. For example, if there are many small grains, then these small grains provide much more surface area per unit volume of space than large grains. Chemical reactions may occur on these surfaces. So, if there are lots of small grains, then surface

Dust in Galaxies
By David A. Williams and Cesare Cecchi-Pestellini
© David A. Williams and Cesare Cecchi-Pestellini 2020
Published by the Royal Society of Chemistry, www.rsc.org

chemistry is encouraged, while the opposite is true if dust grains are mainly large (that is, comparable to the wavelength of visible light).

As we have seen earlier, small grains tend to cause extinction at short wavelengths, that is, in the ultraviolet, and this is exactly the type of radiation that is particularly damaging to interstellar molecules. So the ISEC, through the dust grain size distribution, affects chemistry in interstellar clouds quite significantly. If the size distribution favours small grains then interstellar molecules may be better protected from the damaging effect of ultraviolet radiation, but if the material of interstellar dust is predominantly in large grains, then molecules are more vulnerable to damage by ultraviolet starlight.

All the various methods we discussed in Chapter 2 to find the chemical composition of dust grains suggested similar compositions. In diffuse clouds, through which starlight passes without too much extinction, we concluded that the dust grains are likely to be composed mainly of silicates and carbons, and of some small dust grains that may be considered as large but poorly defined molecules (PAHs) composed mainly of carbon and hydrogen. We also saw that in denser and darker regions than diffuse clouds, these bare grains may be coated with icy mantles composed mainly of frozen water but with some other simple molecules mixed in the water ice (these dirty icy mantles will be explored further in Chapter 8).

The task for this chapter is simple to state, but rather complicated to carry out. Assuming the chemical composition of the dust grain materials as predicted in Chapter 2, can we calculate a theoretical ISEC predicted by these materials and compare it to the observed ISEC? Making this comparison will help us find the information we need. For example, we'll need to assume a range of dust grain sizes, a size distribution describing how the number of dust grains depends on grain size, and a decision about how the dust grain materials are arranged – is each grain formed of one material or several? All these choices we have to make are, in effect, describing a *theoretical model* of interstellar dust. Having made all these assumptions, if we calculate an ISEC that matches the observed ISEC reasonably well, then we can infer that our theoretical model is close to the actual truth; that is, the assumptions we have made for the grains are realistic. If our theoretical model doesn't fit the observations very well, then we simply have to modify our assumptions and make a new calculation and hope that we find a better fit with observations. But even when our model gives a good fit to observations, we can't say that *our* model is the *exact* description. Other models of dust grains, based on a different physical picture, may also be equally satisfactory.

3.1 SCATTERING THEORY

The basic step in attempting to calculate a theoretical ISEC for a collection of dust grains of various sizes and compositions is straightforward but difficult: we want to describe how light interacts with a dust grain. In the simplest case, the beam of light comes from one direction, and the dust grain can be considered as a sphere that is composed of a single material. What we need to know is how much light is scattered by the sphere and how this depends on the angle of scattering, how much is absorbed by the sphere, and how much light is transmitted unchanged (see Figure 3.1). The light that is scattered by the dust grain together with the light that is absorbed by the dust, is the total amount of the light that is lost from the original beam, and this loss represents the extinction caused by the dust grain.

Fortunately, this fundamental problem was solved over one century ago by a German physicist called Gustav Mie. Although the idea is simple and the solution he obtained is mathematically elegant, it is very complicated and so we won't describe it here. The solution is given for particular values of the radius of the dust grain and the wavelength of light, and also depends on the composition of the dust grain. The composition comes into the calculation through the refractive index of the material; the refractive index is something that has to be determined experimentally for the assumed material of the dust grains. That's why we need to know the composition of the dust grains in the interstellar medium.

The basic calculation shows that dust grains of a particular radius produce extinction that peaks at a wavelength approximately comparable to the size of the grain. Therefore, the fact that extinction varies with wavelength (as we discussed in Chapter 1)

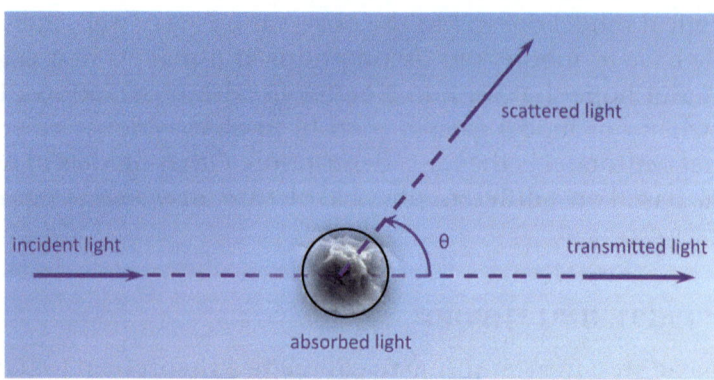

Figure 3.1   The interaction of light with a spherical dust grain. This figure
illustrates schematically the interaction of light from one direction striking the dust grain (this is the *incident* light). The incident light interacts with the grain in various ways. Some light is
scattered, and the amount that is scattered depends on the angle
through which the scattering occurs; some light may be scattered
backwards, and some scattered only slightly. Some of the light
may be *absorbed* by the grain, and the energy in this light helps to
heat the grain. Some light is *transmitted* in the same direction as
the incident light; the transmitted light is obviously weaker than
the incident light. For extinction calculations, we have to add up
these effects for a population of dust grains of different sizes and
materials.

suggests that there is a population of dust grains with a range of
sizes. The next step in the calculation of interstellar extinction is
to add up the contributions to extinction from all the dust grains.
The population of spherical dust grains has radii extending over
a range from a minimum value to a maximum value, and will
have a size distribution in this range. For example, there may
be many more small grains than large, or there might be equal
numbers of dust grains at all radii. The distribution of grain sizes
must also be taken into account. Of course, dust grains of different materials may also be present in the population. If so, allow-
ance must be made not only for the different sizes but also for
the refractive indices corresponding to the different materials in
the dust grain population.

In summary, the information needed to make the calculation of the extinction caused by a population of dust grains
requires quite a lot of information about the population of
dust grains. What is the size range? What is the size distribution within this range? Are the dust grains all of the same

material, or are there dust grains of different materials present? What are the refractive indices of the dust grain materials? The answers to most of these questions (other than the refractive indices, which have to be measured in laboratory experiments) are actually found by "trial and error". We make assumptions about the answers to these questions, we carry out calculations, and we compare the results of the theoretical ISEC with the observational ISECs. If the fit of the theoretical curve to the observational ISEC isn't very good, then we change some of the assumptions made in the hope of obtaining a better fit. We shall discuss in Section 3.2, the choices that have to be made. The procedure is obviously going to guide our choices of the various parameters, and so we obtain the information that we want.

An extension to Gustav Mie's theory allows extinction caused by *composite* dust grains to be calculated. In the simplest form, we suppose that a spherical dust grain is formed by spherical shells of different materials. Then, to make the calculation of extinction, an assumption of the thickness of these shells of different materials has to be made. It is yet another assumption that must be tested by calculation.

The approach outlined here doesn't allow us to calculate the degree of polarization of starlight, because we have assumed that the dust grains are spherically symmetrical. As we described in Chapter 1, to introduce polarization of starlight, the grains need to be asymmetric in some way, such as in the form of discs or of elongated shapes, and partially aligned by some external force. If there is some alignment of asymmetric grains, then looking through such a dust population will cause starlight in one plane of polarization to be extinguished to a greater extent than starlight in another plane. This differential extinction results in the partial polarization of the starlight. Evidently, because polarization of starlight is observed, interstellar dust grains cannot be perfectly spherical. Mie's theory has been extended so that spheres become ellipsoids; that is, solids that in cross-section are elliptical rather than circular, see Figure 3.2.

By choosing to make ellipsoids either very long and narrow, or flat and thin, the method can be extended to shapes such as 'needles' and 'discs', respectively. Alternatively, new methods have been devised that allow the construction of dust grains of any

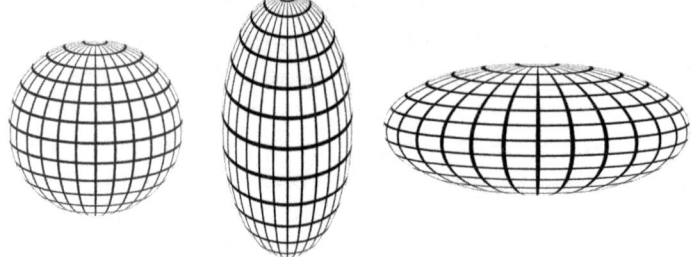

Figure 3.2   Ellipsoidal dust grains. This figure shows schematically how a sphere (left) can be deformed into a spheroid in which one axis of symmetry is very long, approximating a 'needle' shape (centre); or very short, approximating a 'disc' shape (right). Gustav Mie's methods for calculating the scattering of light by a sphere have been extended to allow for ellipsoids.

shape that is desired, from a collection of many smaller spherical dust grains. In fact, most of the calculations are carried out for spherical dust grains. It is normally only when the polarization of starlight is being studied that non-spherical dust grains are considered.

3.2   CHOICES IN A THEORETICAL MODEL OF A POPULATION OF DUST GRAINS

3.2.1   Materials and Their Refractive Indices

Let's adopt as the materials of the dust grains those substances indicated by our discussions in Section 3.1: carbons and PAHs, and silicates. We use the refractive indices of carbons that are well known and can easily be measured in the laboratory.

In fact, carbon exists on Earth in a variety of forms, and soot – a common form of solid carbon – is a mixture of these forms. We shall assume that just two of these forms are important for interstellar dust grains. One of the common forms of carbon is graphitic carbon or *graphite*, a well-known material on Earth. It exists naturally in the form of plane sheets of carbon atoms bound together in hexagonal rings [see Figure 3.3 (left)]. We assume that carbon exists in dust grains partly as pieces of graphite sheets – in other words – what we now call *graphene*. The second form of carbon that we consider is what is called *polymeric* carbon [see Figure 3.3 (right)]. In this form, carbon will

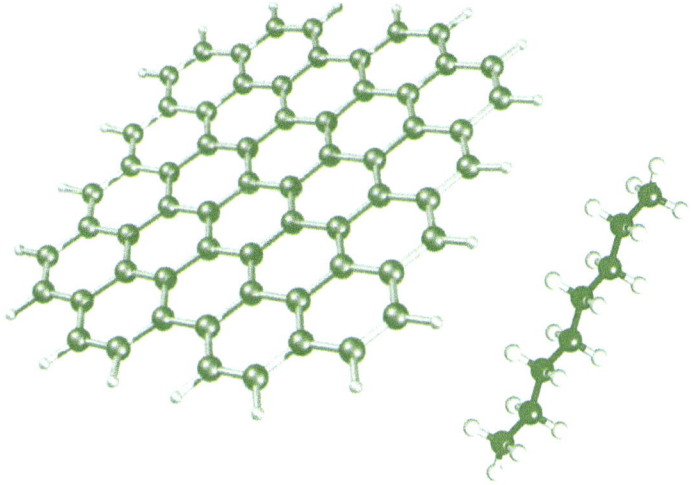

Figure 3.3   Structures of carbon: (left) graphitic carbon, (right) polymeric carbon.

be hydrogen-rich because hydrogen is so much more abundant than carbon in the interstellar medium.

The refractive indices of these forms of carbon are well known. We know that another form of carbon, in the form of free-flying PAH molecules, is probably also present in the interstellar medium. We don't know many of the details for these molecules. But since it seems clear that there will be a variety of sizes and structures of PAHs in the interstellar medium it seems best to calculate the optical properties of a large number of these molecules, and use the average properties in determining the ISEC, rather than trying to measure the optical properties of a few PAHs in the laboratory. The structures of a few PAHs were shown in Figure 2.7.

The refractive indices of various silicates are also well known from laboratory studies. Unfortunately, none of the familiar silicates produce good fits to the observed broader interstellar absorptions in the infrared part of the spectrum at wavelengths near 10 and 20 μm. Terrestrial silicates have absorptions close to these wavelengths but the shapes of the features do not match the observations very well. It seems that the silicates in the interstellar medium may have been altered in some way, perhaps by interacting with radiation in the interstellar medium that silicates don't experience on Earth. It has become customary to modify the optical properties of silicates in such a way that they give a good fit to the two interstellar infrared features with wavelengths near 10 and 20 μm. This theoretical modified silicate is called *astrosilicate*. The problem of not knowing the actual structure of silicates in interstellar space was first noted by a leading expert on interstellar dust, Bruce Draine, and he invented this way out of the difficulty. We should remember that we don't really know what this astrosilicate is; we assume that it is a familiar silicate that has been damaged to some extent by radiation in the interstellar medium. Sometimes, even the most careful scientists have to adopt a 'fudge' like this to make progress, and to hope that the problem will be resolved by future work. It's certainly clear that some kind of silicate material is present in stardust (see Chapter 2), even if we can't yet describe its exact structure.

3.2.2   Size Range of Interstellar Dust Grains

Gustav Mie's theory tells us that dust grains provide maximum extinction at wavelengths that are comparable to the size of the grains. So, it is clear that we need a range of dust grain sizes at least from about a micrometre (corresponding to near-infrared wavelengths) down to a tenth of a micrometre or less (corresponding to wavelengths well into the ultraviolet). In some calculations, we use separate size ranges, one for small dust grains and one for large dust grains, rather than one continuous size range for all dust grains.

These limits are free parameters in the computation of the ISEC that we shall make. But there is a 'cost' in having many large dust grains; they use up a lot of the available material, while small dust grains give 'a big bang for a buck', that is, they have a big effect on extinction for the small amount of material required. Therefore, most calculations are made with the maximum size of dust grains no bigger than about one micrometre, so that the amount of material that the calculation requires to be in dust isn't too large and doesn't exceed the amount of silicon available (as we described in Chapter 2). On the other hand, we know that small dust grains create a steeply rising amount of extinction in the ultraviolet, as is often observed to be the case, so calculations are often made with the lower limit on the grain size as small as about a hundredth of a micrometre (*i.e.*, about ten nanometres).

The actual minimum and maximum sizes can be varied in the calculations until the best fit of the theoretical model to the observed ISEC is obtained.

In addition to the contribution to extinction from the dust grains, we must also include the contribution to extinction from free-flying PAHs. These PAH molecules give rise to the absorption and emission features in the near-infrared between about 3 and 11 μm (see Chapter 2), but they also make a direct contribution to the ISEC. The amount of the carbon tied up in the PAHs is also a free parameter that is determined by the best-fitting theoretical model to the observed ISEC.

3.2.3 Size Distribution and Abundances of Interstellar Dust Grains

We know that there are probably many more small dust grains in interstellar space than large grains. But to define our theoretical model of dust grains completely we need to know more precisely what this size distribution should be. Then we can use this information to estimate the total number of carbon, silicon, oxygen, and other atoms required for our theoretical model of interstellar dust. This number must not exceed the number of these atoms available (as we discussed in Chapter 2). The idea of a size distribution is discussed in Box 3.1.

In Chapter 2 we described the source of interplanetary dust as arising from collisions between asteroids in interplanetary space. The size distribution of the asteroids is controlled – at least in part – by collisions between these solids. The shattering and grinding that occurs in these collisions produces a size distribution of asteroids that is strongly skewed to smaller asteroids, and this is confirmed by observations. In fact, there are roughly a thousand times as many 10 m asteroids as 100 m asteroids. This

---

Box 3.1   Size distributions

What do we mean by a *size distribution*? Perhaps a simple example, unrelated to dust grains, may help.

Think of all the types of coin available in the UK, up to one pound sterling. There are coins for 1 penny, 2 pence, 5 pence, 10 pence, 20 pence, 50 pence, and 100 pence (one pound sterling). If we had equal numbers of all these coins, we could call that a flat distribution. Let's suppose that we have 100 of each of these coins; then the total value would be £188.

But if we were running a shop that dealt in low-cost items, we might find that we had many more of the low-value coins. Let's imagine that the number of coins is as in the table below.

Coin	1 Penny	2 Pence	5 Pence	10 Pence	20 Pence	50 Pence	100 Pence
Number	10 000	2000	400	200	15	8	1
Value (pence)	10 000	4000	2000	2000	300	400	100

Then, we would say that this distribution is strongly skewed towards low-value coins. In total, there are 12 624 coins in this new distribution, while in the flat distribution there were only 700 coins, because in the flat distribution there were relatively many high-value coins. But the total value of the coins is the same in each distribution, £188. So, this value can be distributed among the coins in many arbitrary ways.

The analogy with dust grains is obvious. We have a certain amount of material available to form dust grains with a distribution of dust grain sizes, and the total amount of that material is fixed. But we are free to distribute this material in any way we like, for example, in equal amounts for all sizes, or a distribution that favours larger or smaller dust grains. But the distribution we choose affects the optical properties of the dust very strongly. We have to choose a size distribution that predicts extinction to be similar to the observed observational information, the ISECs. What we find is that we need a size distribution that is strongly skewed towards small dust grains.

---

steep size dependence seems to be typical of solid bodies that undergo shattering and grinding in collisions.

This size distribution may also apply to interstellar dust grains. If so, then there should be about a thousand times as many interstellar dust grains of size 10 nm as at 100 nm. Most ISEC calculations are carried out using an even steeper size dependence in which there are about 3000 times as many dust grains of size 10 nm as of size 100 nm.

It is also possible to leave this size dependence as a free parameter and allow it to be determined by the theoretical model of interstellar dust that provides the best fit to the observational ISEC. If this is done, a very steep size distribution is generally found, with many more small grains than large.

3.2.4 Dust Grains: Composites or Separate?

Assuming that the main components of interstellar dust grains are carbon and silicates, how are these two materials distributed? We could imagine that the materials are entirely separate, so that there is one population of carbon dust grains and another of silicate dust grains. In that case, there might be different size ranges and size distributions for each type of dust grain. These parameters would then be determined by finding the best fit of the theoretical model to the observed ISEC. This approach using separate populations of dust grains is a straightforward way to proceed.

Alternatively, perhaps the grains are composites, that is, mixtures of carbons and silicates. A simple way of thinking about this is to imagine a bare silicate dust grain in a diffuse interstellar cloud. In this cloud there are lots of free carbon atoms and ions. In laboratory experiments involving gaseous carbon, it's impossible to keep carbon atoms and ions in the gas from sticking to surfaces. It's highly likely that a similar process occurs in interstellar space, where the silicate dust grains provide the surfaces. If so, over a period of time, a silicate dust grain would become coated with carbon. Since hydrogen is the most abundant element in the gas, we might expect that some of the deposited carbon would react with hydrogen to form hydrocarbon species, and some might be pure carbon, perhaps a bit like pieces of graphite, or graphene. This process would create composite dust grains, see Figure 3.4.

In this figure, we imagine that the carbon is deposited evenly on the silicate cores, so that uniform carbon shells are formed. The amount of carbon in solid form deposited on the silicate cores increases in time. On the other hand, there are processes that tend to destroy dust grains in the interstellar medium, and one of these is when dust grains find themselves in hot gas rather than the usual cold gas. The hot gas might be caused by a shock

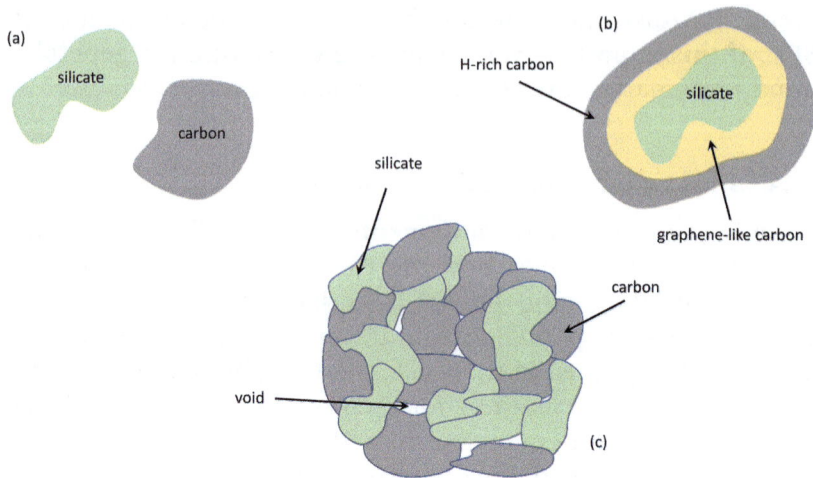

Figure 3.4   How dust grain materials may be arranged in space. In (a), silicates and carbons are separate. In (b), carbons form layers on the surface of the silicate grain. In (c), carbons and silicates agglomerate to form a larger grain with cavities, or voids. This introduces a new component, a vacuum component, in our models of interstellar dust grains.

that arises when two interstellar clouds collide. Whenever the grains are surrounded by hot gas, the outer layers of carbon tend to be removed, some of which may be in the form of PAH molecules. So, these composite grains aren't immutable. They grow when carbon is being deposited; they decay when they are in shocked regions of the interstellar space, where they provide a source of interstellar PAHs. Their story is summarized later, in Figure 5.1.

Calculating the optical properties is more complicated for the composite dust grains illustrated in Figure 3.4b than for the separate model in Figure 3.4a. However, Gustav Mie's method has been extended to deal with composite dust grains like these, and the calculations are routine, if rather difficult.

There is one more matter to be considered. It is possible that dust grains are in fact composed of a number of smaller units packed together [see Figure 3.4c]. If so, then there may be some empty space (*voids*) inside a dust grain, and this space can affect how the grains affect radiation. The simplest way of dealing with this extra component, the vacuum component, is to assume that there is an empty spherical space at the centre of the dust grain.

Alternatively, it's possible to make an assembly of small spherical units packed together to make any shape. Such a model automatically has an amount of empty space within the structure.

3.3   RESULTS

3.3.1   Comparisons of Calculated ISECs with Observational Results

The results of the calculations of ISECs for any lines of sight that have been observed show that excellent theoretical fits can be obtained by the methods described above. These theoretical ISECs closely match the observational data over the entire wavelength range over which extinction measurements are made, from the near-infrared to the far-ultraviolet. Perhaps this excellent agreement between theoretical results and observational results isn't surprising, since – as we have seen – there are many choices that have to be made in these calculations, and each of these choices can be adjusted to improve the fit between calculations and observational results. Nevertheless, these calculations predict the nature of dust grains on any particular line of sight in remarkable detail.

Astronomers sometimes combine results from many lines of sight in the Milky Way galaxy to produce an "average ISEC for the galaxy". The methods outlined above produce a theoretical ISEC that is a very good fit to the observational data. What does this theoretical ISEC predict about the average nature of interstellar dust in the Milky Way? These results are based on the assumption that the grains look like model (c) in Figure 3.4.

In this model, the calculations tell us that the best-fitting results predict a surprisingly high fraction of voids in the silicate core. Almost half the space in the silicate core is empty vacuum between the actual pieces of silicate that don't fit closely together. There is a carbon layer on top of this core that is about one micrometre thick on all grains, both large and small. Nearly all of this carbon is what we call polymeric carbon and only ten percent or so is graphene-like. There are separate ranges of small and large grains, and the size distribution is very strongly skewed towards small grains so that there are several thousand more small grains than large ones. There is also quite a lot of carbon in the form of free-flying PAH molecules. These molecules contribute to extinction over the whole range from the infrared to the far-ultraviolet, and particularly to the prominent "bump" in extinction in the near-ultraviolet.

We can use these results to work out how much silicon and carbon are required in this model of dust grains. It turns out that all the available silicon is required and less than half of the available carbon is needed, and that a large fraction of the carbon is in the form of PAHs. Of course, we mustn't take these results too seriously, because the average ISEC includes a variety of ISECs corresponding to a variety of conditions. But the predictive power of the calculations for particular lines of sight is clearly very strong.

3.3.2   Implications

These models contain at least some of the truth about interstellar dust grains. From studies of ISECs along many lines of sight in the Milky Way galaxy and in other galaxies we can draw the following conclusions:

- It is possible to make appropriate ISECs for dust grain populations made of the materials suggested in Chapter 2.
- Both small grains and large grains are required, and these may be separate populations. It isn't necessary to have one continuous population of dust grains from small to large.
- A steep size distribution is required, with many more small dust grains than large ones.
- Interstellar dust grains may be of composite materials or separate populations of materials.
- Theoretical ISECs can be obtained using all the silicon that is available, but much less than the total available amount of carbon.

3.3.3   But Isn't Dust Changing All the Time in the Interstellar Medium?

We'll discuss more generally in Chapters 4 and 5 how dust grains are formed and destroyed. But, yes, the picture of dust grains in the interstellar medium that we have described in this chapter makes clear that the dust grains are changing all the time. Carbon is being deposited on silicate cores, and the solid carbon changes under the action of starlight from being a hydrogen-rich polymer type of carbon to a carbon that is more like graphene. Eventually, the dust grains find themselves in a hot gas caused by a high-speed collision between interstellar clouds, and the hot gas and collisions between grains erode the carbon layer and eject carbon molecules – including PAHs – into the interstellar gas.

In this description, the dust grains in the interstellar medium are indeed changing all the time. That means that the ISECs associated with these evolving dust grains are also changing. Does that make it difficult to fit the theoretical ISECs to the observations, as we have described in this chapter? Fortunately, the answer is "no". The timescales over which these changes in the ISECs occur are very long in diffuse clouds, typically millions of years. So, the results of the fitting of theoretical ISECs to the observed data are still valid. For long periods of time, the ISECs change by rather small amounts, so the data are valid descriptions of the dust in the regions stated.

3.4 CONCLUSION

As a result of the discussions in Chapters 2 and 3, we have the information that we need to explore the various roles of dust grains in the interstellar medium, and we'll do that in Chapters 6–10 of this book. Before we do that, however, we'll complete the story of interstellar dust grains by considering in more detail where the dust grains come from (Chapter 4), and how they are destroyed (Chapter 5).

Some Old Stars Are "Smoking Like Candles": Is This Where Interstellar Dust Grains Come From?

Many kinds of star vary in brightness, on timescales from days to years to centuries. These variable stars have fascinated astronomers for thousands of years. The first star that was identified (early in the modern astronomical era, in 1638) as a *variable* star is now called Mira; it has a brightness cycle lasting eleven months. Another star, now called R Coronae Borealis, was seen in 1783 to diminish in brightness over a period of a month until it eventually disappeared. It then re-appeared within fewer than two years and its brightness remained constant for about a decade, and then it disappeared again. This star has been behaving like this ever since, but in an irregular fashion. Since those early days, many similar observations have been made of variable stars. There are tens of thousands of variable stars that have been identified in the Milky Way galaxy.

There are many possible causes for stars to vary in brightness. For example, if one star is gravitationally bound to another,

Dust in Galaxies
By David A. Williams and Cesare Cecchi-Pestellini
© David A. Williams and Cesare Cecchi-Pestellini 2020
Published by the Royal Society of Chemistry, www.rsc.org

forming a *binary* star, then when one star passes in front of the other (as seen from Earth) the brightness of the binary pair diminishes and then recovers when they separate. Some stars pulsate naturally, changing their size and temperature as they do so, and these changes affect their perceived brightness in the sky. In this chapter, we are particularly interested in the kinds of stars that vary their brightness because they naturally eject matter to form *circumstellar envelopes*. Stars that have envelopes are abundant in the Milky Way and other galaxies. Some stars do this on a regular basis as part of their evolution, and others may do it in a single evolutionary event, such as the explosion that ends the life of a massive star – a supernova. A brief account of stellar evolution is given in Box 4.1.

It was suggested in 1938 that the special case of the star R Coronae Borealis mentioned above (a star known to be relatively rich in carbon and poor in hydrogen) might be understood on the basis of soot formation in the envelope followed by the ejection of the soot into space. A large amount of soot suddenly appearing in the envelope would certainly extinguish the starlight as perceived by an external observer on Earth, and the ejection of the soot into space, and its expansion there, would allow the starlight to emerge once again. The idea that dust can form in an envelope of gas ejected from a star is now thought to apply to a large class of normal and rather cool stars that are approaching the ends of their lives. The observational evidence seems to confirm that these stars are able to form dust in their envelopes. This dust is expelled into space more or less periodically. This picture was introduced to account for observed variations in stellar intensity, and we now carry the idea one stage further and consider the possibility that these stars are important sources of interstellar dust.

Observations also show us that dust can form even in the ejecta from supernovae. These stellar explosions that end the lives of massive stars are so powerful that for a short time a supernova can be as bright as the entire galaxy (encompassing many billions of stars) of which it is a member. A supernova occurred in the neighbouring galaxy to the Milky Way, the so-called Large Magellanic Cloud, and was detected by astronomers in 1987. We might think that such a violent explosion couldn't be possibly a good place to grow dust grains, but we

Box 4.1   Stellar evolution

On human timescales, most stars appear to be immutable. However, stars do change in time, and can change very much during their lives, from when they are formed until they run out of energy. The description of the way stars change with time is called *stellar evolution*.

At the beginning of the last century, the Danish astronomer Einar Hertzsprung, and independently the American Henry Norris Russell noticed that when stars are plotted using the properties of temperature and luminosity, the majority lie in a smooth band. The resulting diagrams, now called *Hertzsprung–Russell* (or HR) diagrams, are one of the most important tools in the study of stellar evolution. How long a star lives and its ultimate fate depends on how much mass there is when it is formed. The initial mass of a star determines the specific evolutionary stages a star will go through, with massive stars evolving more quickly than low-mass stars. Each stage corresponds to a change in the luminosity and temperature of the star, which therefore moves its position in the HR diagram. Thus, we can learn about a star's internal structure and evolutionary stage simply by determining its position on the diagram.

The majority of stars, about 80–90% of the total stellar population and including our Sun, are found along a region called the Main Sequence (see Figure 4.1). Hot stars are very bright, and so the Main Sequence tends to follow a band going from the bottom right of the diagram to the top left. The luminosity of a star also increases with its size, as bigger stars have a larger surface area through which the star's radiation can flow. This explains the presence above the Main Sequence of stars with the same temperature as cooler Main Sequence stars, but having greater surface areas. Stars that have the same luminosity as dimmer Main Sequence stars, but are hotter (lying to the left of them on the HR diagram), have smaller surface areas. These are termed *dwarf* stars. The converse is also true, and these larger stars are termed *giants*.

We can plot a 'track' on an HR diagram that represents how the temperature and luminosity of a star change over time. In other words, the star moves within the diagram, along an evolutionary path, in response to gravitational contraction and the rate and the fuel of nuclear burning. Protostars change in size because they are contracting under gravity, while Main Sequence stars change because they are using up their nuclear fuel. During such an evolution, some stars pass through a phase in which they are not in equilibrium. These stars may pulsate, expanding and shrinking due to internal forces (mainly thermal pressure and gravity). This pulsation drives a change in the luminosity of the star. During this stage, stars cross an area called the Instability Strip on the HR diagram, a narrow, almost vertical region located in a region between mid-sized stars on the Main Sequence and the Giant Branch. Cepheid Variables are an important type of pulsating stars. They are used as distance indicators because the period of their pulsation varies in proportion with their luminosity. Cepheids are near the top of the Instability Strip.

To estimate just how much the luminosity and temperature of a star change as it ages, we must resort to calculations. A model predicts the luminosity and size of the star, and from these values, we can figure out its surface temperature. For instance, a star like our Sun is initially embedded in a cloud of gas and dust and is not visible outside of this

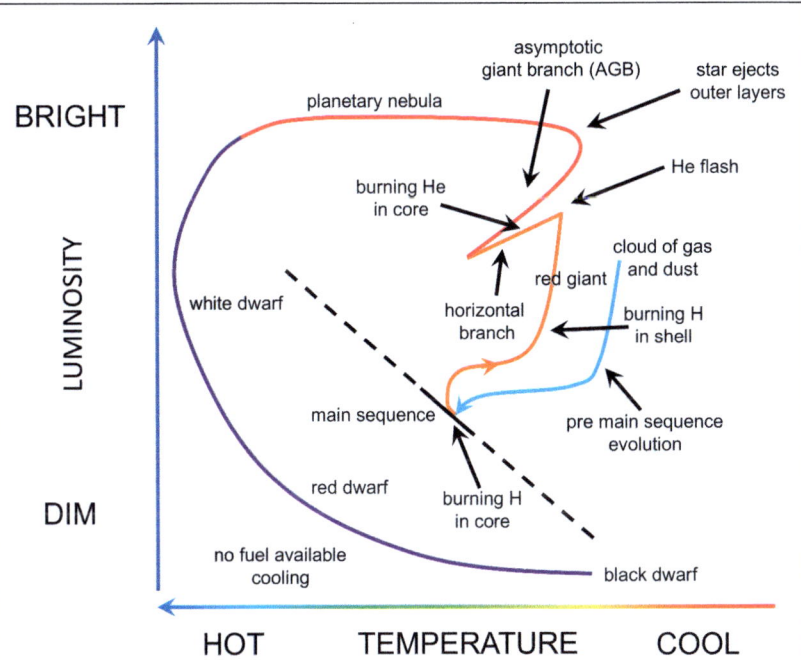

Figure 4.1 The HR diagram. The diagram illustrates the evolution of a solar-mass cloud of gas and dust during its pre-Main Sequence phase (light-blue line) until it arrives on the Main Sequence (black diagonal line) where it burns hydrogen in its core and remains for most of its life. The star then leaves the Main Sequence (following the brown line), passing through various phases (red and purple lines) including the AGB and planetary nebula stages, on its way to eventually becoming a black dwarf.

cloud. At this stage, it is better to call it a *proto-star*. Its internal temperature is about 1 000 000 K, but the outer layers of this proto-star are much cooler, with a temperature of about 3500 K. As the proto-star contracts, the temperature of the outer layers increases, but the net effect is a decrease in its luminosity, so that its representative point on the HR diagram will move down and to the left (10 solar luminosities, 4000 K). This process will continue up to its final location on the HR diagram (0.7 solar luminosities, 4500 K). Finally, when the star is burning hydrogen in its core through fusion reactions and has reached equilibrium, it will lie on the Main Sequence with a temperature of 6000 K and a luminosity of 1 solar luminosity. The time when the star first joins the Main Sequence on the HR diagram is called the Zero Age Main Sequence. Once the star reaches equilibrium, until it exhausts its hydrogen, it stays on the Main Sequence.

(*continued*)

Box 4.1   *(continued)*

A star more massive than the Sun has a hotter core because – thanks to stronger gravity – it can reach higher pressures before they, in turn, can generate enough radiation to slow down the contraction. Thus, more massive stars produce energy at a much faster rate than low-mass stars, and are brighter, but they live for a shorter time. Massive stars stay on the Main Sequence about 10 *million* years, while a star like the Sun has a life time of about 10 *billion* years. The Main Sequence life time of cooler stars may reach the astounding 10 *trillion* years! This time is considerably longer than the current age of the Universe.

What happens when a Main Sequence star exhausts the hydrogen in its core? With no additional fuel in the core, the nuclear fusion converting hydrogen to helium dies out. The core loses the battle against gravity and contracts; as it shrinks, it heats up. This increase in temperature is sufficient to ignite the fusion of hydrogen embedded in the exhausted core. The new, increased radiation pressure actually causes the outer layers of the star to expand. As the gas expands, it cools, which causes the effective temperature to drop. During this stage of expansion, the star will leave the Main Sequence, moving up (higher luminosities) and to the right (lower temperatures) along the so-called Red Giant Branch. The Sun will become a larger but cooler star called a *red giant*. Its size will be 100 times bigger, and the temperature will be about 3500 K.

In the meantime, the helium core – left over of the hydrogen nuclear burning – contracts until its temperature rises over one hundred million degrees. At this point, new nuclear reactions set in converting helium into carbon and oxygen. The star expands again, but not enough to compensate for the increased energy generation, so that the temperature in the core increases without restriction. This occurs because when a gas become super-compressed (the technical term is *degenerate*) it behaves more like a solid. If a normal gas is heated, it expands. However, the pressure in degenerate gas does not depend on the temperature. Because the expansion does not compensate, temperatures remain very high and the helium burning occurs quickly and is uncontrolled. This sudden onset of helium core fusion is called the *helium flash*. During the helium flash the stellar luminosity remains constant. Thus, since the temperature is increasing the star moves to the left, roughly horizontally, across the HR diagram, and hence stars in this phase are said to be on the Horizontal Giant Branch.

At the end of the helium burning phase, the core contracts once again, but to no avail. The end products of helium nuclear burning, carbon and oxygen are unable to ignite their nuclear fusion. However, the core contraction generates sufficient heat for the surrounding layer of helium to start fusing, which in turn heats up surrounding unused hydrogen, which also starts to burn. The giant star expands again, reaching a size as large as the orbit of Mars, and the luminosity increases. The star moves along a track that lies above and roughly parallel to the Red Giant region, that for this reason is called Asymptotic Giant Branch (AGB). In this phase, dust grains are formed in the upper atmospheric layers of the star. Eventually, a wind develops in the star's envelope, blowing the outer layers into space. When the envelope of the star is nearly gone the star becomes a planetary nebula, consisting of an expanding, glowing shell of ionized gas. These objects owe their names to a misunderstanding of William Herschel's, who about 200 years ago called them so because they appeared to be round like planets.

The exposed, remnant core that ionized the planetary nebula material is basically an extremely hot, dense sphere of carbon and oxygen, a so-called *white dwarf* with a surface temperature of about 10 000 K. No more nuclear fusion is possible, and no further gravitational collapse can occur, so energy generation ceases. The star relic radiates its energy steadily, and eventually slowly fades from view.

This description is relative to stars with masses in the range (roughly) 0.8–10 solar masses. Higher mass stars are unlikely to survive as red super giants; instead they will destroy themselves as supernovae (see Section 4.2).

would be wrong! A small amount of dust had already formed in the 1987 supernova within one year, and a quarter of a century later this amount had grown enormously so that about half a solar mass of dust had been created from atoms that – prior to the explosion – were inside the star.

These huge stellar explosions – supernovae – are now considered to be important sources of dust in the interstellar medium, along with variable stars. Also, the less dramatic explosions known as novae, which probably recur, produce dust. The physical conditions in supernovae and novae are different from each other and both are rather more extreme than in the cool envelopes of aging stars, yet dust is able to form. Although the details of dust formation are not accurately known, Nature itself seems to have no problem in creating dust in many kinds of situation, including in huge outflows from very hot and massive stars. Dust formation is also found in inconvenient places such as in high-vacuum laboratory apparatus, where soot deposits are definitely not wanted! Evidently, dust formation is a common feature in our Universe.

In this chapter, we'll describe dust formation in several kinds of circumstellar situation. These are the main sources of dust for the interstellar medium. What happens to this dust when it arrives in the interstellar medium is the subject of the next chapter.

4.1 DUST FORMATION IN THE EXPANDING ENVELOPES OF COOL STARS

It has been known for a very long time that some stars appear to vary in brightness. In some cases the variation is regular, while in others it is erratic. In Figure 4.2 we show some so-called *light curves* – that is, how the intensities of variable stars at visible wavelengths vary with time – for several of these variable stars.

The case of R Coronae Borealis is rather special, as there aren't many similar stars in the Milky Way. But there is a very large class of stars that seem to have similar variability as they approach the end of their lives. These stars are of low-to-medium mass (up to about ten times as massive as the Sun), and as they reach their last stages they have used up much of their hydrogen fuel. They have developed extended cool envelopes (that is, cooler than the star's atmosphere, but with temperatures of about a few thousand K), and these envelopes are chemically different from the Sun. Because of their advanced age they are relatively poor in

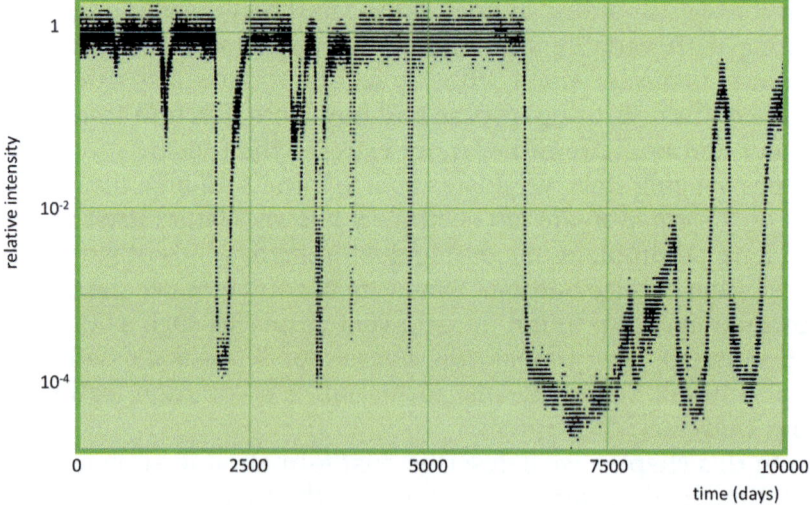

Figure 4.2 Light curves showing intensity changes for the variable star R Coronae Borealis. The intensity of this variable star is observed to fall by a factor of more than a thousand and recover abruptly in irregular periods of months or years. The time interval shown in the figure is 27 years. Data from the American Association of Variable Star Observers.

hydrogen but much richer than the Sun in the elements such as carbon, oxygen, and some metals. All of these elements may be important in dust formation. Some of these stars have atmospheres that have more carbon than oxygen, while others have more oxygen than carbon (like the Sun). The types of dust grains that can be formed are affected by the carbon-to-oxygen ratio. Carbon-rich stars mainly produce sooty grains, while oxygen-rich stars mainly produce silicates.

The relatively cool envelopes around these stars aren't like interstellar clouds, such as the diffuse clouds and the dark clouds that we have mentioned so far in this book. These stellar envelopes are very much denser; in fact, close to the star's surface they are billions of times denser than dark clouds and hundreds of billions of times denser than diffuse clouds. But the density does fall off rapidly further away from the star. We call these envelopes *cool*, but that's relative to the star. The envelopes are cooler than the stellar atmosphere, and they continue to cool as they drift away from the star. But they are much hotter than the interstellar gas in diffuse clouds and dark clouds.

These cool and rich envelopes are good sites for dust formation and this process is certainly the cause of the variability that is observed. If a variable star produces an amount of dust rather quickly, the star's brightness will appear to diminish quite abruptly. Then this dusty envelope drifts away from the star into interstellar space and as it does so it expands. The amount of extinction through this expanding region then diminishes and the star's brightness begins to recover. That's the picture that astronomers believe is happening in this class of variable stars. This idea is discussed a little further in Box 4.2.

There is plenty of observational evidence that dust-formation episodes of short duration do occur in cool stellar envelopes, and that these dusty zones drift away from the star, so that several of these zones can be seen. We see some examples in Figure 4.4. Note that in these images the dust is detected by its emission in the infrared rather than by extinction in the visible region.

In Figure 4.4a, the object called IRC+10216 (astronomers aren't always good with names) is a carbon-rich star surrounded by a thick dusty envelope. In this image we can see that the dust

Box 4.2   Extinction through an expanding envelope

Think of it this way: imagine a cube of gas and dust, of side L cm. This is part of the cool circumstellar envelope surrounding a cool star. This envelope drifts into interstellar space, and expands. Figure 4.3 illustrates this situation.

  Suppose that the number of dust grains per unit volume is n per cubic centimetre. Then the extinction through the cube depends on the number of dust grains along a line of sight through the cube. This is nL dust grains per square centimetre. Now suppose that the cube expands in volume as it moves away from the star so that each side becomes 10 times greater and the volume is 1000 times bigger. But the total number of dust grains in the cube remains the same, so that the number of dust grains per cubic centimetre is $n/1000$; the side of the cube is $10L$, so the number of dust grains along the line of sight through the cube is now $(n/1000) \times (10L)$, or $nL/100$. Therefore, the extinction (which is proportional to the number of grains along the line of sight) is now only 1% of what it was before the change in the size of the cube. Evidently, very large changes in extinction are possible if formation of dust switches on and off rather quickly so that the dust layer then drifts away from the star before the next dust-formation period begins.

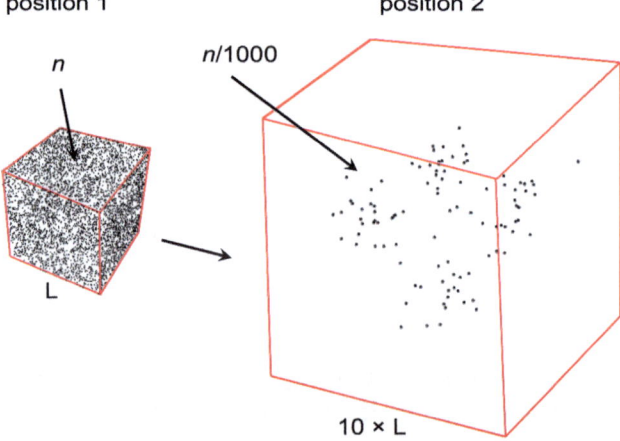

position 1   position 2

n $n/1000$

L

$10 \times L$

Figure 4.3   The expanding envelope of gas and dust. All the gas and dust in a cube at Position 1 in the outflow from a cool star is at a later time contained in the larger cube at Position 2. The number of grains in the cube at Position 2 is the same as at Position 1, but the volume is 1000 times greater, so the number of grains per unit volume at Position 2 is $n/1000$. The amount of extinction at Position 1 is proportional to nL, and at Position 2 is $(n/1000)$ $(10L)$, or $nL/100$. Thus, the amount of extinction provided by this parcel of dust at Position 2 is only 1% of what it was at Position 1.

is formed in periodic events that appear as individual shells. The Ring Nebula [Figure 4.4b] is an object that is at a later stage of evolution in which the central star that had generated a dusty envelope has become a very hot low-mass star. This hot star creates a bubble of hot gas that expands and collides with the cold envelope that is still drifting out into space. The rather beautiful objects such as the Ring Nebula are known as *planetary nebulae* – a very misleading name. In fact, planetary nebulae represent the final short-lived stage of the expansion of cool circumstellar shells into interstellar space. Figure 4.4 shows a few more examples of planetary nebulae showing successive episodes of dust formation in cool envelopes.

We can conclude from the observational information that dust-formation episodes do occur in cool envelopes around these old stars, that these episodes are distinct and don't last very long so that the dust produced in the outflowing envelopes causes the light received by an observer on Earth to vary in an extinction event followed by a recovery. The physical conditions under which dust formation must occur are summarized in Figure 4.5.

Observations show that the outflows accelerate during dust formation, from about a kilometre per second at the beginning of the dust-formation zone to about ten times that velocity at the edge of the zone. The temperature of the envelope is probably too hot for dust-formation in the atmosphere of the star, but falls steadily in the dust-formation zone. The gas density is also lower in the dust-formation zone than in the star's atmosphere, and falls steadily in the dust-formation zone to values that are about a million times the densities in diffuse clouds. The chemistry in the atmosphere and outflow produces simple molecules, but the products become more complicated molecules in the dust-formation zone. Dust formation depends on the formation of nuclei, on which atoms and molecules can be deposited.

Formation of the nuclei is the limiting step. The grains don't grow unless there are nuclei present on which, for example, carbon atoms and molecules can stick and form a sooty particle. Eventually, the dust grains flow out of the formation zone, continue to move away from the star, and finally enter the interstellar medium.

(a)   (b)

(c)   (d)

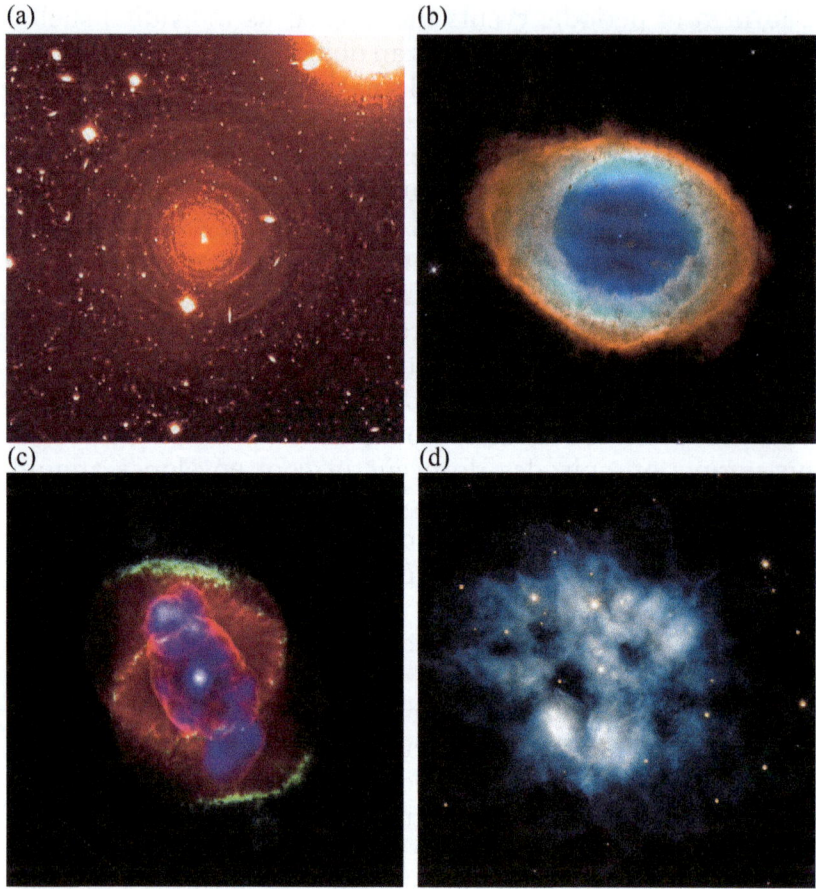

Figure 4.4   Examples of episodic dust formation. (a) Many episodes of dust emission can be seen in this false-colour infrared image of IRC+10216, a carbon star with a thick dusty envelope (credit: I. C. Leao *et al.*, *Astron. Astrophys.*, **455**, 187, 2006, reproduced with permission © ESO). (b) Ejected stellar material is seen in this image of the planetary nebula – the Ring Nebula (credit: NASA, ESA, and C. Robert O'Dell, Vanderbilt University). (c) More complex modes of ejection are seen in the Cat's Eye planetary nebula [credits: NASA/X-ray: Y. Chu (UIUC) *et al.*, Optical: J. P. Harrington, K. J. Borkowski (UMD), Composite: Z. Levay (STScI)]. (d) Ejected material in the planetary nebula NGC 2452 (credit: ESA/Hubble & NASA; Acknowledgements: Luca Limatola, Budenau Cosmin Mirel).

Detailed calculations of the chemistry that occurs in the high-density, high-temperature outflowing envelope gas support the general picture outlined here. For an oxygen-rich gas, the first materials to form solids, at the highest temperatures and

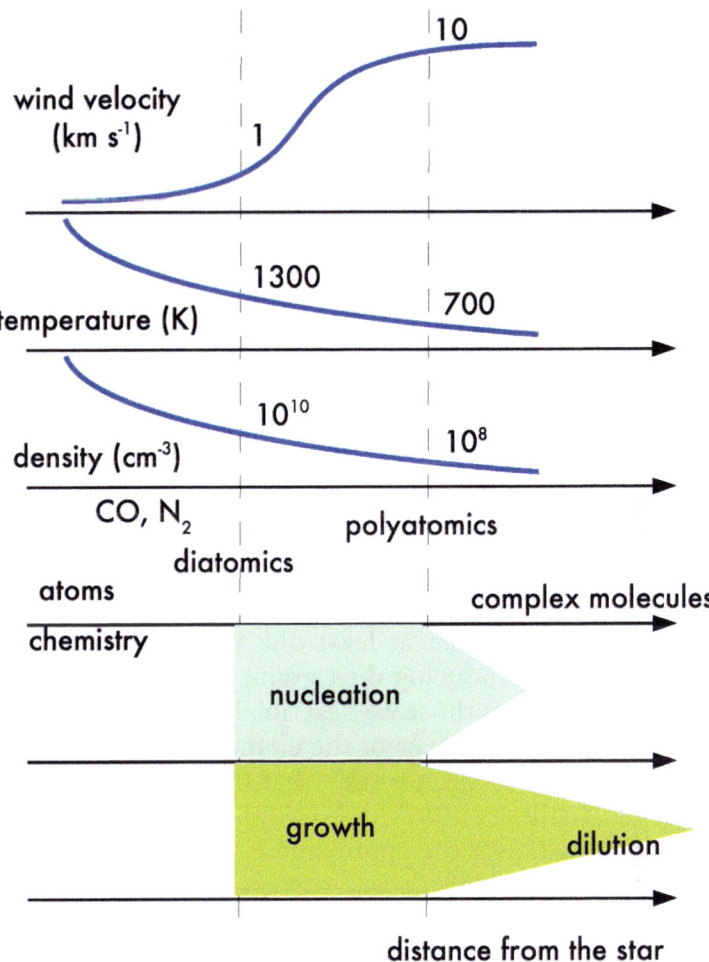

Figure 4.5 A schematic diagram showing the conditions necessary for the formation of dust grains in the outflowing envelopes of cool stars (Reproduced from D. A. Williams and C. Cecchi-Pestellini, *The Chemistry of Cosmic Dust*, 2016, with permission from the Royal Society of Chemistry).

densities, are some fairly obscure oxides that are normally of rather minor abundance such as corundum (Al_2O_3), gehlenite ($Ca_2Al_2SiO_7$), titanium oxide (TiO_2), perovskite ($CaTiO_3$), and zirconium oxide (ZrO_2). These substances aren't abundant enough to supply the materials of interstellar dust, but they can act as nucleation centres on which more abundant species that form later in the expansion can condense. These more abundant species, arriving later in the cooler and less-dense gas, are familiar silicates: forsterite (Mg_2SiO_4) and enstatite ($MgSiO_3$). Forsterite, for example, may be formed by various surface reactions whose net effect may be written as:

$$SiO + 2Mg + 3H_2O \rightarrow Mg_2SiO_4(s) + 3H_2$$

where SiO represents silicon monoxide formed in the envelope, Mg is atomic magnesium in the gas, H_2O is a water molecule in the gas, and (s) indicates formation on a surface.

In carbon-rich envelopes, the chemistry at all temperatures in the outflow is dominated by carbon species, and the most abundant species is the hydrocarbon, acetylene, C_2H_2. This molecule is capable of feeding a network of reactions that generates a large variety of carbon molecules and solids, that is, soot-containing graphene-type structures bonded by simpler carbon molecules.

Therefore, it seems that at least one large class of variable stars is capable of producing dust grains that have a chemical composition similar to those we described in Chapter 2. It's also relevant that the abundances of the elements involved in these dust–formation processes are such that the typical size of dust grains formed in these outflowing envelopes is about a few tenths of a micrometre. This is the approximate size of the large grains in the interstellar clouds.

4.2   DUST FORMATION IN REALLY BIG BANGS: SUPERNOVAE

A supernova is the explosion of a star that occurs at the end of its life, when much of the hydrogen that fuelled the star has been turned into heavier elements such as oxygen, carbon, nitrogen, and metals. The exploding star may be a single star (a Type II supernova) or in a binary (a Type I supernova). Supernovae are astonishingly energetic. Growth to maximum intensity of a supernova and the following decay normally occurs in a few weeks or months, and at peak intensity a supernova may be as bright as the entire galaxy of which the exploding star is a member. In the period of maximum intensity, a supernova will emit roughly as much energy as the Sun will emit in its entire existence of more than ten billion years. Why do some stars end their lives in this dramatic way? Box 4.3 gives a brief description of how this happens.

A supernova explosion drives most of the material of an old massive star into the surrounding space at astonishing speeds of up to about $30\,000$ km s^{-1}, sweeping up and shocking a shell of interstellar matter that becomes very hot indeed, and

---

Box 4.3   Anatomy of a stellar explosion: the death of a single massive star

Just as the force of gravity on Earth attracts everything including human beings towards the Earth's centre, a massive star generates a huge force of gravity that pulls all the matter inside the star towards its centre. But the star doesn't collapse, because that inward force is resisted by the pressure of the hot gas that makes up the star, and throughout almost all the life of the star the pull of gravity and the pressure of the hot gas are in balance, and so the star is stable. The gas pressure is maintained by the heat generated in the thermonuclear reactions that power the star. These reactions begin with the conversion of hydrogen nuclei to helium nuclei, then helium nuclei are converted to carbon nuclei, and a sequence of reactions goes through stages in which the nuclei of neon, oxygen, silicon, and iron are formed in reactions all of which give out energy that maintains the high temperature and high pressure inside the star.

The reaction that forms iron nuclei is the end of this sequence. Iron nuclei are the most tightly bound of all nuclei, so the next stages to convert iron nuclei into even more massive nuclei actually use up energy, and the temperature and pressure begin to fall. Then all the nuclear reactions that only work at high temperature switch off, and there is a massive catastrophe: the temperature and pressure inside the star suddenly fall abruptly. Without internal pressure support, the star collapses under its own gravity: it implodes.

The collapsing star smashes into the core of the star and rebounds, sending shock waves into the outer layers of the star, ejecting them into space at high speed. These layers contain the products of all the nuclear reactions that had previously energized the star. So the surrounding space is seeded with all the elements from helium up to iron, together with any unburned hydrogen in those layers. What remains of the massive star is either a neutron star or a black hole.

The death of stars is vitally important for galaxies and for the Universe. Supernovae populate the space around them with the heavy elements formed from the original fuel – hydrogen. These elements contribute to the gas in interstellar space, and are the elements from which a complex chemistry can emerge and from which dust grains will form. As we'll see in later chapters, the dust and chemistry eventually lead to quite complex molecules related to biology.

In fact, it's fair to say that we owe our very existence to supernovae that created the elements enabling chemistry and biology to develop.

therefore emits radiation at all wavelengths from X-rays to radio waves. This structure is called a *supernova remnant*, and it has a long life time – in contrast to the short duration of the explosion itself.

The last supernova directly detected (by naked eye observations) in the Milky Way was seen in 1604, but it is likely that many supernovae have occurred in our galaxy and have not been detected. The earliest known detection was in 185 AD, and a particularly bright supernova was observed in 1006 AD, followed soon after by the detection in 1054 AD of a supernova that now appears on the sky as a supernova remnant known as the Crab Nebula (shown in Figure 4.6). A supernova detection was also made in 1572. The failure to detect many supernovae in the Milky Way has arisen because the Milky Way contains a lot of dust, and the extinction it causes makes it difficult to see objects across the galaxy, even if they are very bright indeed. It is thought likely that several supernovae per century occur in the Milky Way, but of a possible total of, say, fifty supernovae in two millennia only the five mentioned above have been detected. However, several supernova remnants have been detected in the Milky Way. In fact, it's much easier for us to detect supernovae in other galaxies. Astronomers had a ringside seat in 1987 when a supernova was detected in the Milky Way's immediate neighbouring galaxy, the Large Magellanic Cloud. Some images of supernovae and supernova remnants are shown in Figure 4.6.

Surely the very hot gas created in the explosion cannot be an environment in which tiny solid dust grains are created? And yet, the observations of supernovae tell us that this is so. Dust is formed, and quite quickly. It now seems that supernovae are significant contributors to interstellar dust in the Milky Way galaxy. The best opportunity we have had to observe dust formation in supernovae is with SN 1987A which, as we mentioned in the introduction to this chapter, has produced significant amounts of dust in the few decades since the explosion.

Theoretical studies suggest generally that about a tenth of a solar mass of dust should be produced by a supernova. This dust should be composed of a variety of materials including

(a)

(b)

(c)

(d)

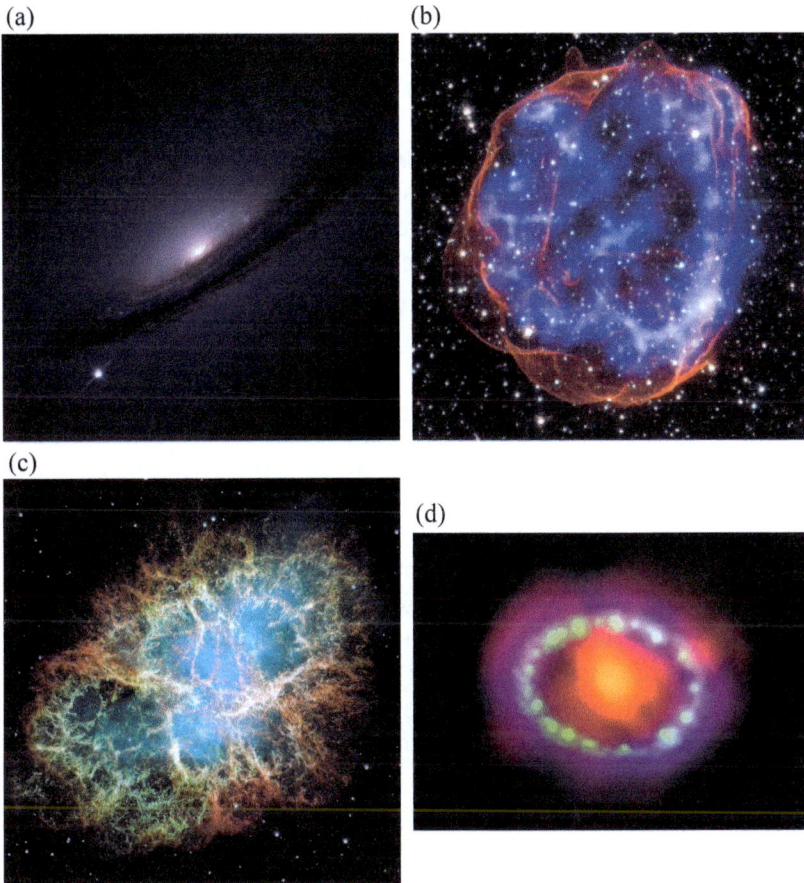

Figure 4.6 Images of some supernovae and supernova remnants. (a) The supernova SN 1994D is of comparable brightness to its neighbouring galaxy NGC 4526 (credit: NASA/ESA, The Hubble Key Project Team and The High-z supernova Search Team). (b) An image of the supernova remnant SNR 0519690 is created from X-ray emission from very hot gas (in blue, Chandra observatory) and in visible light (Hubble Space Telescope) (credit: X-ray: NASA/CXC/Rutgers/J. Hughes; Optical: NASA/STScI). (c) The supernova remnant of a star that exploded in 1054 AD; known as the Crab Nebula [credit: NASA, ESA, J. Hester and A. Loll (Arizona State University)]. (d) A recent image of the supernova SN1987a [(credit: ALMA: ESO/NAOJ/A. Angelich/Hubble: NASA, ESA, R. Kirschner (Harvard–Smithsonian CfA/Gordon and Betty Moore Foundation) and P. Challis (Harvard-Smithsonian CfA) Chandra: NASA/CXC Penn State/K. Frank *et al.*)].

carbons, silicates, and oxides. The dust grains produced are expected to have sizes extending over a wider range than in interstellar dust, and there are predicted to be many more small dust grains than large.

Of course, not all the dust formed after the supernova survives the journey into the interstellar medium. The dust grains must pass through the shock created when the material ejected from the star hits the interstellar gas. Inevitably, some destruction of the newly formed dust grains occurs then, by erosion in the hot gas.

4.3   DUST IN FAIRLY BIG STELLAR BANGS: NOVAE

Novae are stars whose brightness is observed to increase abruptly by thousands of times, and then to fade over some weeks or months. In the days of naked eye astronomy, a nova appeared to be a new star, not previously known, and so it was called a *nova stella*, or new star. Novae occur much more frequently than supernovae, and it is estimated that about fifty novae occur in the Milky Way each year, although only about ten of these are detected. Some novae recur at intervals of decades, while others are single events. Many novae are detected in external galaxies.

Novae occur in close stellar binaries where two stars of modest mass grow old at different rates. Suppose that one star has used up its fuel and reaches the end of its life as a white dwarf, while the other star is still evolving and has an extended atmosphere. The strong gravity of the white dwarf may drag some of that hydrogen-rich atmosphere from its partner and compress it into a thin layer of the surface of the white dwarf. This layer of hydrogen is heated by the white dwarf to such high temperatures that thermonuclear reactions occur, forming a range of heavy elements. The entire layer explodes as a result of the energy produced in these reactions. The explosion injects these new elements into space and the explosion is seen as a nova. The ejecta from the explosion ultimately mix with surrounding interstellar gas. An example of the interaction of the explosion of a nova with the interstellar gas is shown in Figure 4.7.

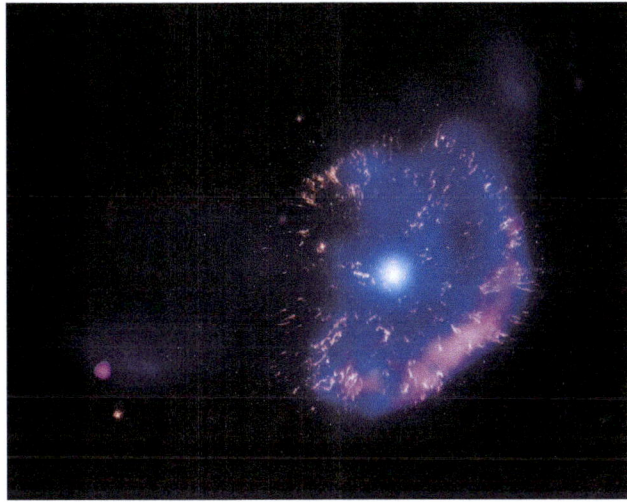

Figure 4.7 The interaction of the ejecta from a nova explosion with the interstellar gas. Regular explosions on the surface of the star eject
material into the surrounding space (credit: X-ray (blue): NASA/
CXC/RIKEN/D. Takei *et al.*; Optical (yellow): NASA/STScI; Radio
(pink): NRAO/VLA[/i][/b]).

In some novae, the early decline in brightness is followed by
a rapid transition to almost the pre-explosion brightness level.
Later on, the brightness recovers onto a path that matches the
initial decline. Infrared observations show that this transition is
caused by the formation of dust in the ejecta. Figure 4.8 shows
schematically how the optical brightness of a dust-forming nova
evolves.

Therefore, it's clear that some novae are capable of forming
dust from the elements that were formed on the white dwarf
and were ejected by the explosion into space. Observations
show that dust formed in a single nova can include both carbon and silicate dusts, unlike dust formed in cool stellar envelopes, which show a preference for one of these types. Other
types of dust in novae have also been detected by infrared
emissions, including silicon carbide (SiC), involving both carbon and silicon. Nova dust grows up to about a micrometre in
size, somewhat larger than the size of dust inferred from measurements of interstellar extinction (Chapter 3). But this dust,

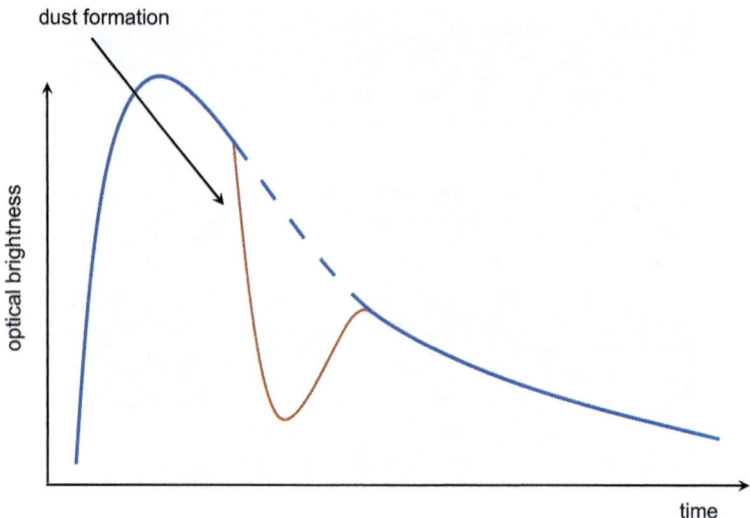

dust formation

optical brightness

time

Figure 4.8 A schematic diagram illustrating how the optical brightness of a dust-forming nova varies in time. The brightness detected by an optical telescope on Earth rises abruptly from its normal stellar value to a brightness hundreds of times greater. Then it fades gradually, but falls abruptly as dust is formed and obscures the optical emission. Much of the radiation is absorbed by the dust and the energy of this radiation is emitted from the dust in the infrared part of the spectrum. As the dusty envelope expands, the extinction caused by the envelope diminishes, and the optical brightness increases, eventually regaining the original path of its gradual decay.

like the dust from cool stellar envelopes and from supernovae, mixes with interstellar gas and is modified there (as we'll see in Chapter 5).

4.4 THE SOURCES OF INTERSTELLAR DUST

Obviously, there are various places in the Milky Way and in other galaxies where dust is formed and injected into interstellar space. Some of these places seem rather hostile to the formation of tiny solid particles, yet the observations show clearly that dust is present, and theoretical studies are able to explain how the apparent difficulties of high temperatures and strong radiation fields are overcome, so that dust grains can form. We have concentrated on three types of sources of dust grains: cool stellar envelopes, supernovae, and novae, but there are others – apparently even more unlikely – in which dust is seen to form. We haven't considered them because their overall contribution is rather small. It is impressive that Nature somehow finds a way to form dust when the abundances of atoms of carbon, oxygen, silicon, and various metals are high enough. Dust seems to be an inevitable outcome of the life cycle of stars.

Each of the main contributors to dust formation in the Milky Way galaxy injects dust at a certain average rate, and in Chapter 5 we'll compare this injection rate with the rate at which dust grains are destroyed to see if these formation and destruction processes can account for the average abundance of dust grains that we can see in the Milky Way. The injection rates take account of the number of sources of dust (that is, how many cool stellar envelopes, supernovae and novae there are, on average, in the Milky Way) and are given in terms of units of solar masses of dust, per unit area of the plane of the Milky Way, per unit time, and we'll take the unit area to be a square of the plane of the Milky Way one light year across, and unit time to be ten million years. The results are shown in Table 4.1.

Table 4.1 shows that cool stellar envelopes and supernovae are the most important contributors of dust into the Milky Way galaxy. Carbon dust covers a range of carbon products, and in addition to silicates a variety of other types of dust, such as oxides, are

Table 4.1 Approximate injection rates of the main types of dust into the Milky Way galaxy, in units of solar masses of dust, per light year squared, per ten million years.

Source	Carbon dust	Silicate dust
Cool stellar envelopes, C-rich	3	—
Cool stellar envelopes, O-rich	—	5
Novae	0.3	0.03
Type I supernovae	0.3	2
Type II supernovae	2	10

present. Therefore, some stars at the end of their lives do indeed "smoke like candles" and provide dust grains that enable the ongoing formation of stars and planets and may seed the Universe with molecules related to biology. Of course, the first stars in the Universe didn't have the benefit of dust to assist in their formation. They were quite unlike the stars in the modern Universe.

CHAPTER 5

What Happens to Stardust in Interstellar Space?

Dust grains injected into interstellar space from exploding stars and cool stellar envelopes – we call these grains *stardust* – don't last forever. In fact, dust grains have a rather eventful life, and eventually come to a bad end: they are destroyed. There are various interstellar processes that tend to destroy dust grains. Some of these processes destroy dust grains instantaneously and turn them back into their constituent atoms; some re-distribute matter and alter the grain size distribution; and others change the physical properties of the dust material. If the dust is embedded in interstellar gas that collapses to form a new star, then that dust will be destroyed inside the star, once the star 'switches on', because of the very high temperatures in stellar interiors. We'll meet star formation later, in Chapter 9. In this chapter we'll discuss only those processes that affect dust in the interstellar medium, after the dust is formed in stellar explosions and cool stellar envelopes.

The effects of dust that we observe in interstellar space aren't necessarily the immediate consequences of dust grains emerging from exploding stars and circumstellar envelopes. For example,

Dust in Galaxies
By David A. Williams and Cesare Cecchi-Pestellini
© David A. Williams and Cesare Cecchi-Pestellini 2020
Published by the Royal Society of Chemistry, www.rsc.org

when we detect interstellar extinction caused by dust grains along a particular line of sight, there may be, in that collection of grains, some young dust grains fresh from their formation. There will certainly be some middle-aged grains that have been significantly modified by events in interstellar space, and inevitably there will be some old grains that have been heavily processed by events in interstellar space and which may bear little resemblance to newly formed dust grains. Interstellar space operates like a giant mixing machine, mixing gas and dust from various locations together. What is driving this mixing machine? It's basically due to turbulence from various sources. One important origin of turbulence is the overall rotation of the galaxy, which induces a spin in any parcel of gas. This tends to mix interstellar matter on large scales. We'll discuss turbulence in more detail later on, see Box 9.2.

Another source of turbulence arises in stellar explosions, particularly supernovae explosions. As we saw in the previous chapter, these explosions drive very high-speed winds into interstellar space, and these winds can travel very far if the gas density is quite low. But where these high-speed winds run into interstellar clouds they penetrate and shock the cloud. The effect on the gas in the shocked region is to convert some of the energy of the fast wind into heat, and the temperature of the gas may rise to very high values. This very hot gas spells death for interstellar dust!

5.1 DUST DESTRUCTION IN SHOCKS

Not much can happen to a dust grain sitting quietly in a low-temperature gas. But interstellar space is often a rather violent place. We have seen in the previous chapter that supernovae and novae eject matter into space at high velocities. If this high-speed gas runs into pre-existing gas, then a shock is set up, and the effect of this is to convert part of the energy of motion of the ejected matter into heat. The consequent temperature rise can be very large and in the case of a supernova wind colliding with interstellar gas the gas may be so hot that it emits not only in the visible and ultraviolet but in the X-ray part of the spectrum. In Figure 4.6 we showed some images of supernova remnants, and some of the emission from the remnants is in the X-ray region.

The X-ray emission shows precisely where a high-speed shock has been formed by the impact of the supernova wind on existing interstellar material.

In fact, a large fraction of rather empty interstellar space is affected by supernova winds and heated by them to high temperatures. Where the winds impact on interstellar clouds things get rather complicated. There are various possibilities that can occur. First, atoms and ions either from the wind itself or from the hot gas may strike a dust grain and knock out atoms or molecules of the dust material. This process is called *sputtering*, and it gradually erodes the surface of the dust. This affects small and large dust grains alike. Second, fast grains can collide with slower grains. Collisions like these, if energetic enough, may cause *shattering* of one or both of the grains, creating a large number of smaller grains in the impact. Obviously, this changes the size distribution. The final possibility is that if the energy in the collision between two grains is great enough, then one or both of the grains may be reduced immediately to atoms. This process is called *catastrophic shattering*; it is indeed a catastrophe for a grain that is destroyed in this way. It simply ceases to exist and the materials of which it was made are returned to the gas as atoms.

Supernovae aren't the only cause of shocks in interstellar space, but they are responsible for the most violent shocks. Interstellar shocks also arise in other ways. For example, collisions between interstellar clouds can occur. The speeds involved in collisions like these are much less than those from matter ejected by supernovae, but still capable of generating high temperatures. Cloud–cloud collisions occur more frequently than supernovae. Interstellar space is a turbulent place, and collisions between gases occur with speeds from just above the speed of sound (about a kilometre per second in gas of modest temperatures) up to as much as a thousand kilometres per second. We conclude that dust grains in interstellar clouds live a fairly quiet life punctuated by occasional disruptive events – any one of which may be terminal!

It's possible to make calculations about the outcome of these events. Recent results suggest that shattering in collisions between grains begins even at relatively low speeds, but large speeds are required to make this an effective process. The size of fragments produced in shattering becomes smaller as the speed of impact increases, but the number of fragments becomes larger. However, catastrophic, shattering doesn't play a significant role until the impact speeds are very high. Sputtering is important for returning the materials of dust to the gas, but in terms of the destruction of dust grains, sputtering isn't as important as shattering.

All these results depend on information about the dust grains, and unfortunately that information is poorly known. It is safe to make a general conclusion that large grains emerging from cool stellar envelopes or supernovae will be converted by shattering to generate dust grain size distributions in which there are many more small grains than large. The size distribution may be similar to the distribution inferred from modelling interstellar extinction, as we described in Chapter 3. But such a conclusion depends on knowing something about how well the dust grains can resist shattering. At the level of grain structure, we really don't have good information about the porosity of the material in dust grains, or whether grains are composed of smaller pieces bound together in some way and how weak or strong this binding might be, or whether the solid is – instead – quite homogeneous. All of these factors, and more, will affect the ability of dust grains to withstand shattering.

Even at the atomic level within the dust, things are unclear. Consider the case of carbon materials. Some carbons can be eroded in the shocked gas formed in these events while others are more resistant. It's well known that carbon in the form of diamond is a very hard material that can resist sputtering and shattering much better than other forms of carbon such as graphite or polymeric carbon. Therefore, the nature of carbon material may be changed during the lifetime of a carbon grain in interstellar space, because hard materials may survive where softer materials are eroded or shattered.

One of the consequences of the erosion of dust grains is that some softer carbons in the dust grains may be eroded and carbon put back into the gas. Then it is free to stick to grains again, and since it can do this in the presence of abundant hydrogen it seems likely that carbon is deposited on

hard dust grains as carbon rich in hydrogen, that is, a hydrocarbon layer on top of a core of harder material, say, a silicate
grain. A diagram outlining this sequence of events is sketched
in Figure 5.1.

But even this re-arrangement of matter in the dust grains
isn't the end of the story. Hydrocarbons are well known to be
affected by ultraviolet radiation. So we have to consider what
happens to hydrocarbon solids in interstellar space, where
radiation from hot stars is strong in the ultraviolet part of the
spectrum.

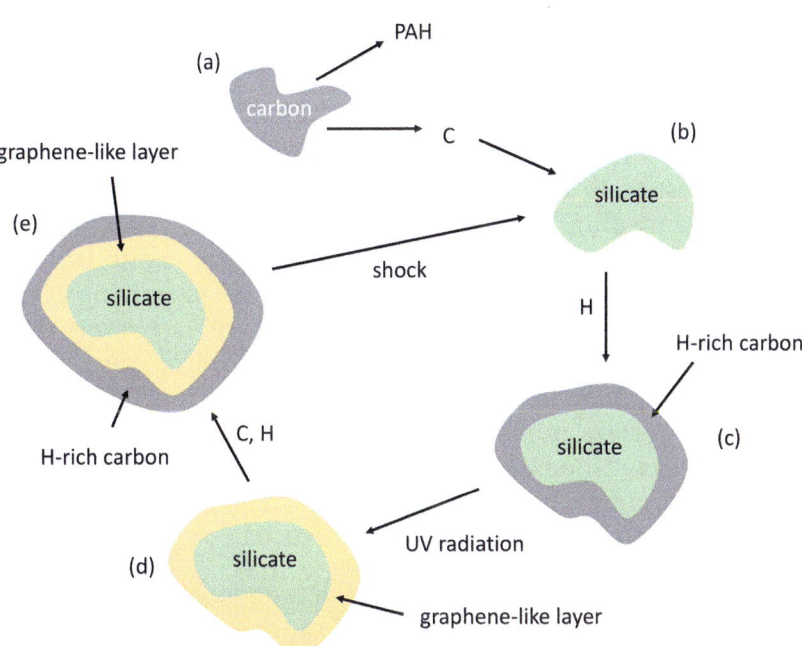

Figure 5.1 Re-arrangement of carbon in interstellar dust grains. (a) A carbon
grain is eroded, releasing carbon atoms and PAH fragments to the
gas. (b) Carbon atoms strike a bare silicate grain, and are retained
at the surface. (c) In the H-rich interstellar medium, the carbon
layer on the silicate becomes H-rich. (d) The strong ultraviolet
(UV) interstellar radiation field drives out hydrogen from the surface carbon layer on the silicate (*photodarkening*), creating a type
of carbon that is similar to graphene. (e) Carbon atoms from the
interstellar gas continue to be deposited and create an outer layer
of H-rich carbon. An interstellar shock may remove both carbon
layers, restoring the bare silicate grain.

5.2   IRRADIATION OF DUST IN INTERSTELLAR SPACE

In laboratory experiments, when solid carbon that is rich in hydrogen – that is, a hydrocarbon solid – is exposed to ultraviolet radiation the hydrogen tends to be driven out. The solid then becomes hydrogen-poor and undergoes a re-structuring. Carbon bonds that were attached to hydrogen now find themselves without a 'partner' and a further re-arrangement takes place in the carbon, with carbon atoms bonding with other carbon atoms, rather than hydrogen atoms. The end result is that the carbon contains regions that are like graphene, containing hexagonal rings of carbon atoms. This structural change is accompanied by a change in appearance: the material appears darker than it was. The process is called *photodarkening*. Evidently, we may expect photodarkening to occur in the interstellar medium, where the ultraviolet radiation field is strong. We can use information from experiments in the laboratory about the efficiency of photodarkening to estimate how long this change takes to occur in the interstellar medium.

There are several processes happening at the same time in interstellar space. Carbon atoms from the gas are arriving at the surface of a dust grain and being stuck there as hydrocarbon material. Ultraviolet starlight is turning this hydrocarbon material into a graphene-like solid layer on the surface of a dust grain. Then more carbon atoms arrive, making new hydrocarbons – and the process continues. So the picture of the carbon layer is that the most recent carbon to arrive is a hydrogen-rich outer layer, and underneath that is the material that has been on the dust grain for the longer time. The inner layer has been converted to graphene-like material. This is the explanation for the structure of composite dust grains that we described in Section 3.2.4, and sketched in Figure 3.4b.

Eventually, dust grains like these will enter a region of shocked gas in interstellar space. The dust grains are disrupted and the carbon layers will be removed. The pieces of graphene-like material enter the gas as PAH molecules. The cycle of carbon into and out of grains was illustrated in Figure 5.1.

Eventually, a dust grain may enter a shocked region in which the temperature is so high that it is reduced to atoms. That's the end of that grain's existence! The life cycle that probably began with the birth of a dust grain in the cool envelope of an old star of modest mass or in the explosion of an old and massive star ends in cremation. Then the cycle repeats. New dust grains are formed and ejected into space; they evolve, changing their optical properties, and eventually die, to be replaced by the next generation of dust grains.

An interesting aspect of this picture is that the dust grains should change their extinction properties during the evolution, and this is confirmed by calculations. Figure 5.2 shows the ISECs for dust grains evolving as described here, in which the hydrogen number density in the cloud takes several different values.

These ISECs are also changing in time. What astronomers actually see when they measure extinction is an average of instantaneous ISECs corresponding to different stages of evolution and to different gas densities along the line of sight. So the properties of dust grains that are deduced from observations also correspond to some kind of averaging. We shouldn't take physical properties of dust deduced from these extinction averages too seriously. Perhaps it's better to think of the deduced properties more as a helpful guide than an actual 'prescription'.

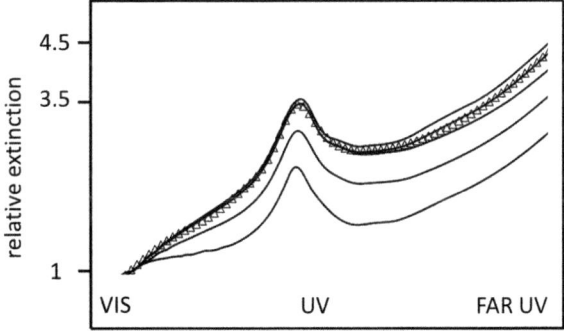

Figure 5.2 Instantaneous normalized ISECs for several interstellar clouds with different densities. The figure shows ISECs computed after two million years of evolution. The different curves represent the ISECs for a gas of H atoms with number densities 30, 100, 300, 1000, and 3000 atoms per cm^3, from top to bottom. The triangles show the average ISEC for the Milky Way galaxy.

5.3 FAST PARTICLES

Interstellar space is pervaded by fast particles. Some of these are called *cosmic rays*, although that is a poor description. These 'rays' are in fact particles moving with very high speed, and the particles are the nuclei of atoms. The most abundant particles are hydrogen nuclei, but nuclei from many other atoms are also present. These ultrafast particles have an important role to play in making molecules in the interstellar gas, and we'll discuss that role in more detail in Chapter 6. This role depends on cosmic rays interacting with and ionizing gaseous atoms and molecules. However, cosmic rays also pass through dust grains, and it is a curiosity of quantum mechanics – the way that experiments tell us to describe how atomic matter behaves – that the faster the particle, the smaller its interaction with the dust. It seems rather counter-intuitive; one would expect the more energetic the interaction, the more complex the inter-action might be. But it isn't so, and these very fast particles – cosmic rays – don't interact very strongly with the material of the dust grains. They do, however, give a little energy to grains as they pass through. The energy heats up the grains a small amount. Small grains heat up more, while large grains heat up less. Small grains can even become warm enough to eject into the interstellar gas some of the atoms and molecules that are weakly bound to their surfaces. Larger grains may not be able to eject these weakly attached travellers, and if more such molecules continue to accumulate on the surface, then the dust grains will grow a layer of dirty ice. These ice layers are of great importance in astronomy and astrobiology, as we'll see in Chapter 8.

Quantum mechanics also tells us that the interaction of nuclei with the grain will be much stronger if the nuclei have slower speeds. It's as if a slow nucleus has more time to become involved with the atoms in a dust grain. Think of a jogger running through a park; she has no time to talk to passers-by, but someone else just strolling along through the park may often stop to chat with others. Slower atoms may also have more 'conversations', that is, more interac-tions with atoms in a dust grain. There are also much slower nuclei than cosmic rays in interstellar space, and these are also mostly hydrogen nuclei. These slower nuclei can affect the structure of silicate grains. Experiments show that silicates called *olivines*, involving both magnesium and iron, are affected by the impact of slower nuclei. Magnesium seems to be expelled from the bulk of a silicate and enrich the surface. If this is happening to interstellar dust grains, then these irradiated grain materials aren't like any silicates that we know on Earth. So it's perhaps not surprising that – as we saw in Chapter 3 (Section 3.2.1) Bruce Draine had to invent the material that he called *astrosilicate* to have the correct properties to account for interstellar extinction, but which is unlike any known silicate on Earth.

5.4 THERMAL EFFECTS

Interstellar silicates are found to be *amorphous* rather than *crystalline*. Amorphous solids are those without any long-range order (like glass), whereas crystalline materials are defined by their regular structure (like common salt). Crystalline solids have absorption spectra that are clearly distinct from those of amorphous solids formed of the same materials. Interstellar silicates lack the distinctive spectrum of crystalline materials, and are therefore amorphous.

On the other hand, collected silicate dust grains from comets are found to be crystalline, rather than amorphous. This seems to be a puzzle: since interstellar dust is found to be amorphous, and cometary dust is simply interstellar dust that has been through a star-forming event (as we'll discuss further in Chapter 9), it is important to understand how silicate grains that were originally amorphous (as in interstellar dust) become crystalline (as in cometary dust). Figure 5.3 shows jets of dust emerging from the comet 67P Churyumov–Gerasimenko, imaged during the Rosetta mission to that comet during 2014–2016. This cometary dust is mainly crystalline. How does dust that was amorphous (in interstellar dust) become crystalline (in cometary dust)?

A good way to turn an amorphous material into a crystalline one (assuming that a crystalline form exists) is to heat the material until it becomes liquid, and then allow it to cool slowly. If the cooling is too fast then the material will solidify

Figure 5.3   Dust jets on comet 67P Churyumov–Gerasimenko imaged during the Rosetta mission. Reproduced from http://www.esa.int/spaceinimages/Images/2015/02/Comet_on_6_February_2015_NavCam under the terms of the CC BY 3.0 Share-Alike IGO licence, https://creativecommons.org/licenses/by-sa/3.0/igo/.

with its molecules 'jumbled up' in an irregular fashion as it is in the liquid state, that is, in an amorphous state; but with slow cooling the crystallinity emerges. Evidently, in slow cooling the atoms and molecules making up the crystal have time to find the most stable positions, which are the positions that guarantee crystallinity.

But a comet is basically a dusty snowball, so how can crystals form in a body that must be at a very low temperature? What appears to be happening is this: before the comet was formed, some dust passed near the newly-forming star and was strongly heated. Then the dust moved further away and cooled slowly so that the dust became crystalline; then the comet-forming stage occurred later. Situations in which dust grains are so strongly heated that crystallization can occur are very rare. Therefore, most of the dust in interstellar space is amorphous.

5.5 CONCLUSIONS ON THE LIFE CYCLE OF DUST

Is the story of the life cycle of interstellar dust complete? In qualitative terms, it certainly seems to be complete. We know the main sources of dust in interstellar space. Dust is formed in some special events such as stellar explosions and cool stellar envelopes, and is ejected into interstellar space. While in interstellar space, dust is affected by various processes that alter the size distribution of the grains and make re-arrangements of the matter in the grains. The interstellar dust created by these processes generates the extinction of starlight that astronomers observe. Eventually, dust grains meet their end when they pass into a high-speed shock set up by a supernova. Indeed, some grains that are born in a supernova may also die in a supernova shock wave.

The picture seems qualitatively convincing. Does it work quantitatively? Do the processes of formation and destruction of dust grains described here generate the amount of dust that is observed in the Milky Way galaxy? At the moment, we aren't sure. The processes of destruction seem to be too fast, or else the processes of formation seem to be too slow to account for the amount of dust observed in the Milky Way. But estimating global figures for the galaxy, for both formation and destruction, is very difficult. It may be that both estimates will be revised in future work. Perhaps dust grains are more robust than is currently believed to be the case, or it may be that the formation of dust is actually rather more effective than we currently understand.

CHAPTER 6

Doing Chemistry in the Dark: How Interstellar Dust Leads to Molecular Complexity in Interstellar Gas

6.1 MOLECULES AS PROBES OF INTERSTELLAR GAS

In Chapters 1–3 we showed that the primary effect of interstellar dust on interstellar gas clouds is that the intensity of radiation penetrating into the interior of the clouds is reduced. Box 6.1 describes the various contributors to the radiation field in interstellar space.

As we've seen in Chapter 1, dust grains in a cloud scatter and absorb starlight, so that the intensity of starlight at the centre of the cloud is reduced, or even totally extinguished, compared to the intensity at the cloud's edge. This reduction (or extinction) in intensity applies not only to visible light but also to infrared and ultraviolet radiation. In a typical diffuse cloud, the intensity of visible light in the interior of the cloud may be reduced to about 60% of the external value, depending on how much dust is in

Dust in Galaxies
By David A. Williams and Cesare Cecchi-Pestellini
© David A. Williams and Cesare Cecchi-Pestellini 2020
Published by the Royal Society of Chemistry, www.rsc.org

Box 6.1   What is the interstellar radiation field?

The Milky Way galaxy contains a huge number of stars, very roughly about a hundred billion. Other galaxies may be larger and have more stars, or smaller and have fewer stars (see Chapter 1 and Figure 1.1). All the stars in a galaxy are radiating huge amounts of energy into space. The sum of all this stellar radiation is an important part of the interstellar radiation field, and controls the chemistry that can occur in interstellar clouds. Apart from starlight, there are several other important components of the interstellar radiation field, too, and we'll mention those below.

Starlight

We are most familiar with radiation from the Sun. Solar radiation warms planet Earth and makes life on our planet possible. The Sun's radiation peaks in the green part of the visible spectrum, a wavelength of about 500 nm; our eyes are sensitive to the whole visible spectrum of wavelengths from less than 400 to more than 700 nm; we see this whole integrated range as white light.

The Milky Way, like all galaxies, contains stars that range in mass from much smaller than the Sun to much larger. Smaller stars are very numerous, and they are cooler than the Sun. They are much less powerful than the Sun and they radiate at longer wavelengths, mainly in the infrared. On the other hand, stars that are more massive than the Sun are rarer in the Milky Way, but they are hotter and very much more powerful. For example, a star that has a temperature about three times as high as the Sun's, radiates mainly in the ultraviolet, but is perhaps a hundred times more powerful than the Sun.

So the interstellar radiation field created by stars ranges from the infrared through the visible into the ultraviolet. It is the sum of contributions from very numerous but relatively weak low-mass stars to very rare and very powerful massive stars. In the case of the Milky Way, the rarity of the massive stars is overcome by their huge power, and the interstellar radiation field created by all the stars is dominated by the massive stars.

In the Milky Way, this interstellar radiation field from stellar radiation is, therefore, mainly in the ultraviolet part of the spectrum. This turns out to be a very important part of the spectrum for interstellar molecules, because almost all molecules are destroyed by ultraviolet radiation. We'll discuss in Chapter 7 the particular case of the destruction of molecular hydrogen by the interstellar radiation field. But it's generally true that molecules tend to be destroyed by ultraviolet radiation, because this radiation often carries enough energy to break the bonds between the atoms that make molecules, or to ionize a molecule, making it easier to destroy in other ways. For example, the CH molecule (methylidyne) may be affected by UV radiation as follows:

$$CH + UV \text{ radiation} \rightarrow C + H$$

$$CH + UV \text{ radiation} \rightarrow CH^+ + e$$

(*continued*)

Box 6.1   (*continued*)

Breaking bonds successively can destroy larger species quite effectively, for example, the common interstellar molecule CH_3OH (methanol) can be destroyed by UV radiation in this way, removing atoms step-by-step:

$$CH_3OH \rightarrow CH_3O \rightarrow CH_2O \rightarrow HCO \rightarrow CO \rightarrow C + O$$

unless other reactions help to restore the molecule.

If a molecule is unshielded from the interstellar radiation field in the Milky Way, it is likely to be destroyed by radiation in about one thousand years. That's a long time for us, but rather a short timescale for the formation of interstellar molecules, so astronomers generally don't find molecules in the unshielded parts of the interstellar medium. The ability of dust grains to shield molecules from the destructive effects of the interstellar radiation field is the subject of this chapter.

Other Components of the Interstellar Radiation Field

What happens to starlight? A lot of the energy in starlight is eventually absorbed by the dust grains, and heats them. So there is a flow of energy into the grains, and this is balanced by a flow out, because the grains themselves radiate, even though they are very cold indeed. The temperature of large grains in the unshielded interstellar medium is typically about ten degrees above absolute zero and smaller grains are a bit warmer, but even at these temperatures the grains are radiating. This radiation has wavelengths in the hundreds of micrometres. The total amount of energy in this radiation is a bit smaller than – but comparable to – the total amount of energy in the starlight. So in energy terms it is very important on the galactic scale, but on the microscopic scale it doesn't have a major role to play in interstellar chemistry. We won't need to refer to it again in our discussion of interstellar chemistry.

There is also another major radiation field in the interstellar medium; it has a total energy that is smaller than (but comparable to) the total energy in the radiation from dust. The radiation peaks at millimetre wavelengths and its origin is cosmological. It is a relic of the Big Bang in which the Universe, according to standard cosmological theory, is supposed to have begun. The Universe – including the radiation within it – has been expanding ever since it began. This *cosmic microwave background radiation* (CMBR) can affect the population of low-lying rotational states of molecules, but is unlikely to play a significant part in interstellar chemistry in the modern era. However, if we were thinking about chemistry during a much earlier stage of the Universe's evolution then this CMBR could be important. We shall not need to refer to the CMBR again. Extending to much longer wavelengths than the millimetre CMBR are radio waves created by motion in ionized interstellar gases. These regions aren't important in interstellar chemistry.

Finally, there are very energetic local events taking place in the interstellar medium, such as explosions called supernovae and novae (see Chapter 4 and Figure 4.6). Events like these generate very high temperatures and are very powerful radiation sources beyond the ultraviolet region, into the X-ray region of the spectrum. However, these events have a local influence near to such sources, rather than a general effect in the interstellar medium. So for regions near to supernovae, it may be important to allow for these local radiation fields, or for other local fields that arise from radioactive atoms ejected from stellar interiors in these explosions. In terms of the chemistry occurring in interstellar clouds, we shall not need to refer again to these radiation fields generated in explosions or by radioactive atoms.

the cloud. In the infrared, the extinction caused by the dust is weaker and in a typical diffuse cloud (which has around a hundred hydrogen atoms per cubic centimetre) the infrared intensity in the interior is only slightly reduced, to about 80% of its external value. However, we have seen that extinction caused by interstellar dust grains in the ultraviolet is much stronger than in the visible. Even in the relatively low-extinction diffuse clouds the ultraviolet radiation intensity may be significantly reduced to about 5–10% of the external intensity.

For clouds that are denser and darker than diffuse clouds, such as Barnard 68 pictured in Figure 1.5, then the extinction is much stronger. B68 isn't unique. Figure 6.1 shows some more images of dark clouds. Clouds like these have about ten thousand hydrogen atoms per cubic centimetre, in the form of hydrogen molecules.

At visible wavelengths, the intensity of the interstellar radiation field in the interior of dark clouds is typically less than one ten thousandth of the external intensity, while the ultraviolet intensity is totally extinguished in the interior. The infrared internal radiation field is also reduced, but only to about 15% of its external value. There are several important points arising from these results.

Firstly, if we want to probe inside these denser clouds, then it will be much easier to do that using radiation at infrared wavelengths, rather than at visible or ultraviolet wavelengths. This conclusion tells us immediately that radiation emitted or absorbed between rotational levels of molecules will be suitable to use, because these rotational transitions in molecules correspond to

(a)   (b)

(c)   (d)

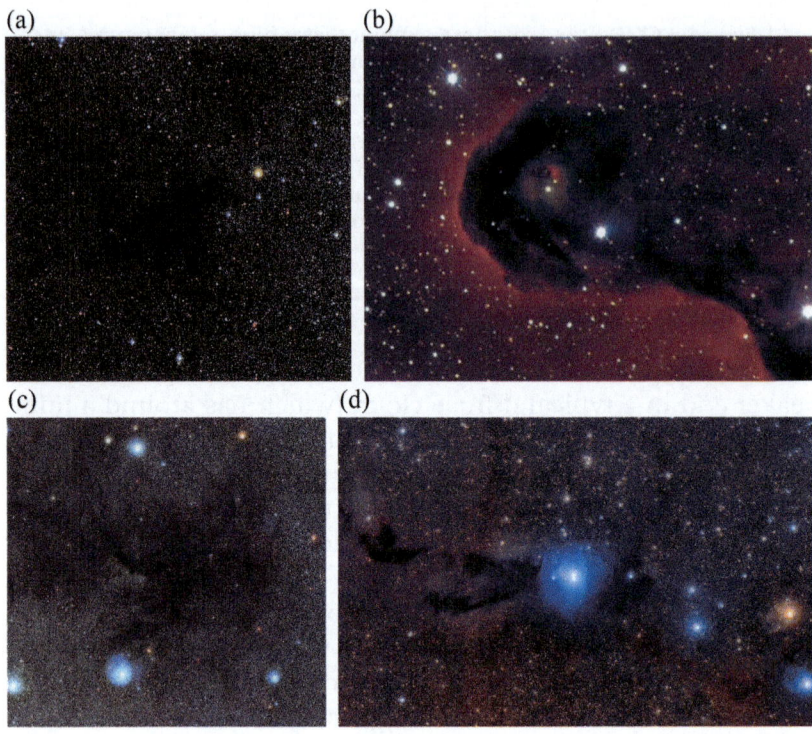

Figure 6.1   Images of dark clouds. (a) The dark nebula LDN 483, located about 700 light years away in the constellation of Serpens; it very effectively obscures the light of many background stars (credit: ESO). (b) VDB 142, also known as the the Elephant Trunk Nebula; it is a spectacular star forming region about 2400 light years distant in the constellation of Cepheus (credit: Arne Henden, US Naval Observatory, Flagstaff; image processed by Al Kelly). (c) The Pipe Nebula, an irregularly shaped dark cloud, part of the Great Dark Horse Nebula (the 'rump' and 'back legs' (credit: ESO/Y. Beletsky). (d) The Lupus 3 region, 600 light years away in the constellation of Scorpio. It is a region of star formation, and a cluster of stars has already emerged (credit: ESO/R. Colombari).

wavelengths typically of millimetres, where the extinction is very low. We discussed these rotational spectra briefly in Chapter 1. Consequently, we need to consider the kinds of molecules that may exist in interstellar clouds; these may be good probes of the conditions in these clouds.

Secondly, ultraviolet radiation is more energetic than visible or infrared radiation, and it is a major contributor to the heating and ionization of gas in interstellar clouds – if it can penetrate them. Since it is effectively excluded from the interiors of dark clouds, these interiors have very low temperatures and very low ionization fractions. The temperature and ionization inside clouds are maintained by other less effective energy sources than the interstellar radiation field. The interiors of dark clouds turn out to be some of the coldest and least ionized regions of space in the Universe. Typically, the temperatures in these regions are only about ten degrees above absolute zero (or 273° below zero Celsius, the temperature at which water freezes on Earth. You can refer to Box 2.1 for a discussion of temperature). The ionization fraction is typically also very low indeed: there is about one electron per hundred million hydrogen atoms inside dark clouds. Diffuse clouds are much warmer than dark clouds, but still very cold by human standards. They are typically about one hundred degrees above absolute zero (or about 173° below zero Celsius), and they are more strongly ionized – there is one electron for about ten thousand hydrogen atoms.

Thirdly, molecules are easily destroyed by ultraviolet radiation. So molecules in diffuse clouds have a difficult time, because the radiation field can penetrate into the cloud. But dark clouds have very little ultraviolet radiation in their interiors, so these are locations where we expect molecules to exist in high abundance. Observations show that this is indeed the case.

In dark clouds, where ultraviolet radiation is excluded, the most effective sources of heat and ionization are the so-called *cosmic rays*. We mentioned in Chapter 5 the effects of cosmic rays on dust grains. These 'rays' are in fact highly energetic particles – mostly the nuclei of hydrogen atoms. The origin of cosmic rays is probably in the supernovae that end the lives of massive stars. The energy of these particles is so great that they can penetrate dark clouds with ease and without significant loss of energy. They pervade the whole of the interstellar space of the Milky Way galaxy and probably much of the space outside our galaxy. As we'll see in Section 6.4, cosmic rays are responsible for initiating a large part of the interstellar chemistry that occurs in the gas in dark clouds.

In this chapter we want to focus on the chemistry in clouds and the molecules that survive because extinction caused by dust grains suppresses the radiation field. We shall see in later chapters that the molecules produced by this chemistry are the crucial components that enable star and planet formation to occur (Chapter 9). Also, large complex molecules can arise in icy mantles on dust grains (Chapter 8) in clouds where the extinction provided by dust grains is high, and some of these complex molecules appear to be related to biological molecules (Chapter 10). So, it's clear that the extinction provided by dust grains is essential for all this interstellar chemistry to be possible.

We'll look (Section 6.3) at the chemistry that can occur in diffuse clouds where interstellar extinction is low and the intensity of the interstellar radiation field is high. Then, in Section 6.4, we'll look at the more extensive chemistry that occurs in denser darker regions (such as B68 or those illustrated in Figure 6.1) where it's clear that the dust grains are doing a very thorough job of excluding the interstellar radiation field from the interior of these dark clouds. But first, in the next section, we'll summarize the basic ingredients available for chemistry occurring in the interstellar gas.

6.2 WHAT ARE THE INGREDIENTS FOR CHEMISTRY IN THE INTERSTELLAR GAS?

The most abundant species is, of course, hydrogen. In diffuse clouds, hydrogen is present in roughly comparable amounts of hydrogen atoms (H) and hydrogen molecules (H_2), while in dark clouds the hydrogen is mostly molecular (H_2). The high abundance of hydrogen compared to other elements means that if something can react with either atomic or molecular hydrogen it will do so in preference to reacting with any other partner. The conversion of hydrogen atoms to hydrogen molecules is rather special and will be described in Chapter 7.

What about the other elements? After hydrogen, the next most abundant element is helium, which doesn't form molecules very easily. In fact, the only detected interstellar molecule containing helium is the helium hydride ion, HeH^+, detected in a shell of gas around an old star. So let's turn to the more reactive species. In Chapter 2, we discussed the amounts of various elements that seem to be missing from the interstellar gas, and we said that it was very likely that these missing elements were still present but locked up in interstellar dust grains. From that information, we then speculated about the chemical nature of the dust grains. From Tables 2.1 and 2.2 we can find the amounts of the various elements remaining in the gas: it is these elements that are available for chemistry in the interstellar gas. The three main species are carbon, nitrogen, and oxygen, with about 91, 62, and 389 atoms, respectively, relative to one million hydrogen atoms, while most of magnesium, silicon and iron are in the dust grains, leaving about an atom of each in the gas (Table 6.1). There are about 10 sulfur atoms per million hydrogen atoms, but less than one atom of sodium and calcium per million hydrogen atoms. Therefore, in order of abundance, the important species for chemistry are oxygen (O), carbon (C), nitrogen (N), sulfur (S), silicon (Si), and magnesium (Mg).

In diffuse clouds some of these elements will be ionized by the radiation field, and these are carbon, sulfur, silicon, and magnesium. This means that these atoms have one (negative) electron removed by the radiation field, so they appear as positive ions C^+, S^+, *etc.* In the interior of dark clouds, these atoms are almost entirely neutral.

This information about the ingredients for interstellar chemistry is summarized in Table 6.1. Note that most of the silicon and magnesium is in the dust grains, so their abundances are only about one atom relative to one million H atoms.

These are the ingredients. What are the recipes? What can we make?

Table 6.1 Main ingredients for chemistry in diffuse and dark clouds.

Element	Main form in diffuse clouds	Main form in dark clouds	Abundance in gas, relative to 1 000 000 atoms
Hydrogen	H, H_2	H_2	—
Oxygen	O	O	389
Carbon	C^+	C	91
Nitrogen	N	N	62
Sulfur	S^+	S	10
Silicon	Si^+	Si	≈1
Magnesium	Mg^+	Mg	≈1

6.3 HOW TO MAKE CHEMISTRY IN THE GAS OF DIFFUSE CLOUDS

Making chemistry at the low temperatures and low densities of diffuse clouds isn't easy. Also, any molecules that are produced are fairly rapidly destroyed by the interstellar radiation field that can penetrate the cloud rather easily, for example:

$$OH + starlight \rightarrow O + H$$

Starlight destroys most molecules quite easily (as we may know from personal experience; think of the damage that sunlight can do to the skin of an over-enthusiastic sunbather).

There are lots of atoms, so an apparently obvious method of making molecules would simply be to collide one atom, say oxygen, with a hydrogen atom so that a hydroxyl molecule is formed:

$$O + H \rightarrow OH$$

and then perhaps another similar reaction would form the water molecule (H_2O). Unfortunately, this simple idea doesn't work! The colliding partners (nearly always) bounce off each other before they can re-arrange themselves to become hitched together as OH (hydroxyl) or eventually through a subsequent reaction as H_2O (water).

An alternative might be for an oxygen atom to collide directly with a hydrogen molecule to produce hydroxyl:

$$O + H_2 \rightarrow OH + H$$

This reaction – a so-called *exchange* reaction between neutral partners – works well at temperatures very much higher than those of diffuse clouds, but is very slow in cold diffuse clouds which, as we said, have temperatures that are only about a hundred degrees above absolute zero.

These simple ideas aren't very encouraging! However, if *ions* rather than neutral atoms are considered, then the results are more promising. For example, in the collision between C^+ ions (created by starlight acting on carbon atoms) and hydrogen molecules, the electric charge of the ion gives an additional attraction to the H_2 molecule that prolongs the collision considerably and helps to form a new species, CH_2^+:

$$C^+ + H_2 \rightarrow CH_2^+$$

This isn't a very efficient reaction; it certainly doesn't happen each time these partners collide, but it is much faster than, say, the reaction of oxygen atoms with molecular hydrogen (which is very slow at low temperatures). The newly formed CH_2^+ molecule may react with another H_2 to form CH_3^+ which attracts a negative electron to form the neutral molecules CH (methylidine) or CH_2 (methylene):

$$CH_3^+ + e \rightarrow CH + H_2, \text{ or } CH_2 + H$$

What about oxygen chemistry? How does oxygen chemistry begin? Although oxygen atoms are mainly neutral, some are ionized by reactions with hydrogen nuclei (H^+) that are created when cosmic rays collide with hydrogen. These O^+ ions open up a new entry route into the chemistry, because they react directly with H_2, unlike their neutral counterparts:

$$O^+ + H_2 \rightarrow OH^+ + H$$

and in subsequent reactions the product OH^+ extracts H atoms from H_2 molecules to form H_3O^+ which combines quickly with an electron to give OH and H_2O. The O^+ reaction is an exchange reaction, but one of the partners is charged. This makes all reactions like these much faster, because the charge helps to keep the reactants together for longer than in the case of neutrals. Exchange reactions like this are called *ion–molecule* reactions.

These entry channels into diffuse cloud chemistry can overlap. For example, the reaction of C^+ with OH is fast and gives CO^+:

$$C^+ + OH \rightarrow CO^+ + H$$

which reacts with H_2 molecules in a fast reaction:

$$CO^+ + H_2 \rightarrow HCO^+ + H$$

In diffuse clouds the ion HCO^+ (the formyl ion) soon meets an electron and is neutralized, splitting off an H atom:

$$HCO^+ + e \rightarrow CO + H$$

forming the important interstellar molecule, carbon monoxide, CO. Evidently, a large part of interstellar chemistry is initiated by cosmic ray ionization of atoms and molecules in the interstellar medium.

Once we can identify a starting point for the chemistry, the network of reactions becomes complicated very quickly. In spite of the difficulty of finding useful entry routes into the chemical network, diffuse cloud chemistry can be extensive. The agreement between calculations based on this chemical network (usually involving hundreds of reactions) and identifications of molecules observed to be present in diffuse clouds is good. A list of some of the molecular species detected in diffuse interstellar clouds is shown in Table 6.2. These molecular species are formed in reactions in the interstellar gas, similar to the reactions described above.

Most of these detected species are fairly simple molecules, containing only two or three molecules. These simple molecules, especially OH, H_2O, and CO, have very important consequences, as we'll see in Chapter 9, where we discuss the formation of stars and planets in interstellar gas clouds. The abundances of these species reach a steady state, that is, a balance between formation and destruction reactions, in about one million years. After that time, the abundances predicted by these calculations don't change in time, unless the physical conditions are changed.

But significant chemical complexity in interstellar clouds doesn't arise from the network of reactions indicated here. For that, we have to turn to dark clouds from which the ultraviolet radiation field is almost entirely excluded. However, it's worth remembering that diffuse clouds do contain PAHs, the large hydrocarbon molecules that contribute to interstellar

Table 6.2 Some molecular species, formed from H, O, C, and N, detected in diffuse clouds.

No. of atoms	Species
2	H_2 (molecular hydrogen), HD (deuterium hydride), CH (methylidyne radical), NH (nitrogen monohydride), OH (hydroxyl radical), C_2 (diatomic carbon), CN (cyanogen radical)
3	CH_2 (methylene radical), NH_2 (amino radical), C_2H (ethynyl radical), HCO^+ (formyl ion)
4	CH_3 (methyl radical), NH_3 (ammonia), H_2CO (formaldehyde)
5	c-C_3H_2 (cyclopropenylidene)

extinction. These are likely to be formed in the erosion of carbon particles. There is also good evidence that some *fullerene* molecules, C_{60}^+ and C_{70}^+ exist in the diffuse clouds, and the corresponding neutral species have also been detected in interstellar space; see Box 6.2. Fullerenes have a characteristic shape (see Figure 6.2) and are sometimes referred to as "cage" molecules.

Box 6.2 What are fullerenes?

Carbon exists naturally in a variety of forms, of which graphite and diamond are well known. In 1985 a remarkable new form of carbon was made by Richard Smalley, Robert Curl, James Heath, Sean O'Brien, and Harold Kroto at Rice University. They made C_{60}, a spherical molecule of sixty carbon atoms bound together in a mixture of hexagons and pentagons. The structure resembles a conventional football; see Figure 6.2. The C_{60} molecule was given the name *buckminsterfullerene*, because the structure is similar to the large geodesic domes built by the architect Buckminster Fuller. Members of this whole family of carbon molecules make take spherical, ellipsoidal, or cylindrical shapes; they are all called *fullerenes*.

   These fullerenes have been detected spectroscopically by their absorption at particular wavelengths in the infrared, corresponding to vibrations in these structures.

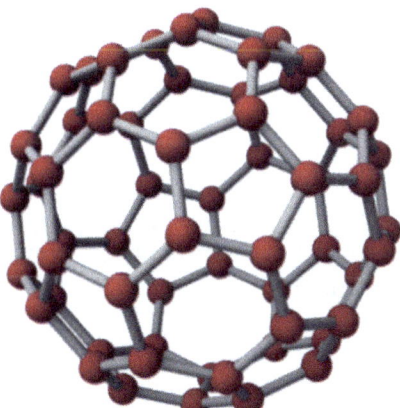

Figure 6.2   Diagram of the fullerene C_{60}. C_{60} is the simplest member of the fullerene family. The structure is a mixture of hexagons and pentagons (as in a soccer ball), with carbon atoms at all the vertices.

It seems likely that fullerene molecules are also formed during the creation and destruction of carbon solids in space, a process we described in Chapter 3.

6.4   HOW TO MAKE CHEMISTRY IN THE GAS OF DARK CLOUDS

As we emphasized earlier, dark clouds are really dark and really cold. This means that reactions between the main forms of the available ingredients listed in Table 6.1 don't occur. Therefore, just as we did for diffuse clouds, we have to identify new ways to start the chemistry in dark clouds. Once it is started, the chemistry becomes very complex very quickly.

In diffuse clouds, we saw that ions C^+ and O^+ were important in starting the chemistry. In dark clouds, however, starlight is excluded so carbon isn't ionized by stellar radiation. The oxygen ions in diffuse clouds were formed in reactions of O atoms with the H^+ ions that were created by cosmic rays interacting with hydrogen. It turns out that cosmic rays (the fast particles pervading interstellar space – see Section 5.3) start the chemistry in dark clouds by ionizing hydrogen molecules, which are by far the most abundant species present:

$$H_2 + \text{cosmic ray} \rightarrow H_2^+ + e + \text{cosmic ray}$$

where e represents an electron; it carries a negative charge. The most likely thing that can happen to the newly formed H_2^+ ion is that it bumps into another hydrogen molecule. The ion captures a hydrogen atom to form a new species, H_3^+, the trihydrogen ion. This unusual triangular molecule has the property of being able to give a hydrogen ion, H^+, to many other atoms and molecules; for example, to oxygen:

$$H_3^+ + O \rightarrow OH^+ + H_2$$

Just as in diffuse clouds, the OH^+ reacts successively with molecular hydrogen, extracting a hydrogen atom on two occasions:

$$OH^+ \rightarrow H_2O^+ \rightarrow H_3O^+$$

but the sequence stops there because O^+ can bind three hydrogen atoms to itself, but no more. So a likely consequence is that the H_3O^+ ions react with an electron to form hydroxyl and water:

$$H_3O^+ + e \rightarrow OH + H_2 \text{ and } H_2O + H$$

where the electrons arise from the ionizations of hydrogen by the cosmic rays. This scheme forms an entry route into interstellar oxygen chemistry.

A similar scheme works for carbon, and produces the molecules CH (methylidine), CH_2 (methylene), CH_3 (methyl), and CH_4 (methane), together with the ions CH^+, CH_2^+ and CH_3^+. This is an entry route into interstellar carbon chemistry. An even more complicated carbon chemistry can obviously develop from these carbon-bearing ingredients. For example, carbon chains can begin to grow by inserting carbon atoms into CH_3^+:

$$C + CH_3^+ \rightarrow C_2H^+ + H_2$$

and in further reactions the C_2H^+ ions can generate C_2 (diatomic carbon), C_2H (ethynyl), and C_2H_2 (acetylene). The potential schemes continue almost indefinitely, and long carbon chains (called *polyynes*) can develop through reactions such as:

$$C_2H + C_2H_2 \rightarrow C_4H_2 + H$$

where C_4H_2 (diacetylene) is a chain of four carbons with a hydrogen at each end. Reactions of carbon chains with nitrogen atoms can create carbon chains with a CN (cyanide group). One of the largest molecular species of this linear type detected in interstellar clouds and created by this kind of chemistry is HC_9N, which has the splendid name cyano-octa-tetra-yne. It is a chain of eight carbon atoms with a hydrogen atom at one end and a cyanide (CN) group at the other.

Once the simple species have been established in the oxygen and carbon entry chemistries, then more complex species can form from reactions between their products; reactions like these proceed much more easily as the complexity increases. For example, when CH_2 and OH are present in the gas, then formaldehyde (H_2CO) can form in a simple exchange reaction:

$$CH_2 + OH \rightarrow H_2CO + H$$

and formic acid (HCOOH) also forms in a simple exchange reaction involving formaldehyde and hydroxyl:

$$H_2CO + OH \rightarrow HCOOH + H$$

Simple schemes involving small species soon lead to chemical complexity. For example, methanol (CH_3OH) may be formed in the reactions involving the association of an ion and a molecule:

$$CH_3^+ + H_2O \rightarrow CH_3OH_2^+$$

followed by a neutralizing of the ion with an electron:

$$CH_3OH_2^+ + e \rightarrow CH_3OH + H$$

Or, in the case of ethanol (C_2H_5OH), a similar process of association:

$$C_2H_4 + H_3O^+ \rightarrow C_2H_5OH_2^+$$

followed by a neutralizing of the ion with an electron:

$$C_2H_5OH_2^+ + e \rightarrow C_2H_5OH + H$$

Schemes such as these may generate chemical complexity of a large number of molecular species in networks of thousands of reactions.

There are some cases that need extra consideration. Nitrogen atoms don't accept H^+ from H_3^+, so the network of entry reactions that applies to oxygen and carbon doesn't apply to nitrogen. However, as the existence of interstellar HC_9N indicates, nitrogen atoms can take part in reactions with many of the species produced in the oxygen and carbon chemistries described above. For example, nitrogen may react with hydroxyl:

$$N + OH \rightarrow NO + H$$

to form the detected species nitric oxide (NO), and with simple hydrocarbons to form cyanides:

$$N + CH_3 \rightarrow HCN + H$$

where HCN is hydrogen cyanide.

Similarly, sulfur doesn't accept H^+ in reactions with H_3^+, but, like nitrogen, has a rich chemistry that is initiated when sulfur atoms react with species created from the oxygen or carbon entry schemes. For example, sulfur atoms can react with OH:

$$S + OH \rightarrow SO + H$$

and this may be followed by:

$$SO + OH \rightarrow SO_2 + H$$

Sulfur monoxide (SO) and sulfur dioxide (SO_2) are detected interstellar species. Sulfur atoms may also react with the products of the carbon entry scheme, for example:

$$S + CH_3 \rightarrow H_2CS + H$$

forming the equivalent of formaldehyde (H_2CO) with sulfur replacing oxygen; this molecule (H_2CS) is called thioformaldehyde, and is a detected interstellar species.

These ideas suggest that molecules with greater and greater complexity can be formed in reactions like these between atoms, ions, and molecules with the products of the entry chemistries for carbon (CH, CH_2, and CH_3) and oxygen (OH and H_2O).

While starlight in diffuse clouds tends to destroy molecules and limits the abundances of molecules, there is no starlight in dark clouds. Does this mean that molecular abundances in dark clouds increase indefinitely? No, it doesn't. There is destruction of molecules, too, in the chemistry; each of the reactions listed above shows that the reactants are being used up to generate more complex products.

However, there is also another agent causing the destruction of gas phase molecules, and it works because of the cosmic rays themselves. Helium is famously non-reactive (that's why it is sometimes called a "noble" gas – it doesn't mix with the common stuff!), but when cosmic rays ionize helium atoms, it becomes a demon with devilish appetite. When helium ions are formed:

$$He + \text{cosmic ray} \rightarrow He^+ + e + \text{cosmic ray}$$

the ions are capable of ripping apart almost any other molecule. For example, helium ions destroy carbon monoxide molecules (which are very strongly bound molecules):

$$He^+ + CO \rightarrow He + C^+ + O$$

Of course, the carbon ions are now available to initiate further chemistry, and some of these ions will even go on to form carbon monoxide again. So, as in the case of the diffuse cloud chemistry, there is a balance to be struck between formation and destruction of molecules in the chemistry occurring in the gas of dark clouds. If this balance can be reached, there is no further change in molecular abundances. However, there is another factor that affects dark cloud chemistry: all the atoms and molecules in the gas may collide with dust grains and may stick on the grains' surfaces, forming a dirty ice. We'll consider such ices and their importance to astrochemistry and astrobiology in Chapter 8. For now, we note that if atoms and molecules do indeed stick to grains (as confirmed by the detection of interstellar ices) then the formation and destruction of molecules in the gas can never be in balance.

The chemistry of dark clouds is much more extensive than that of diffuse clouds. A proper description of the chemistry of dark clouds may need thousands of reactions to be reasonably complete. The number of different molecular species that have been identified in dark clouds is currently over two hundred, and new detections continue to be made each year.

A list of all the molecular species detected in diffuse and dark clouds, star-forming regions and circumstellar envelopes is shown in Table 6.3. The chemical complexity shown in the table is remarkable. It reflects the ability of carbon to generate an enormous chemistry. Some of these species might be familiar from school chemistry or other applications: for example, ethylene glycol, $(CH_2OH)_2$, is an anti-freeze we use in the radiators of our cars; acetic acid, CH_3COOH, is the main ingredient in vinegar, and ethanol C_2H_5OH, is the inebriating component in wine, beer, and whisky.

The number of species actually present in dark clouds and involved in the chemistry must be much larger than two hundred, but not all these species are capable of being identified in space. In some cases, their abundances may be too low for detection. In others, they may not have a transition in a convenient part of the spectrum (like the infrared, or in the millimetre waveband); if they don't, then we can't detect them no matter how abundant they are.

If the substitution of isotopes is also included in the chemistry, such as deuterium (D) for hydrogen (H), or ^{18}O for ^{16}O, or ^{13}C for ^{12}C, then the number of molecular species to be considered increases enormously. For example, the single species CH_4 (methane) is replaced by five species if all the substitutions of D for H are included. The five versions are $^{12}CH_4$, $^{12}CDH_3$, $^{12}CD_2H_2$, $^{12}CD_3H$, and $^{12}CD_4$. A similar five each for the isotopes ^{13}C and ^{14}C would be required if carbon isotope substitutions are also included, making a total of fifteen species instead of only one. The fifteen species require many additional reactions to be included in the reaction set. Evidently,

Table 6.3 Molecular species detected in interstellar space. The molecules are listed in groups corresponding to numbers of atoms, from 2–13. The order within a group gives the order of discovery. In addition to these species, the fullerenes C_{60}, C_{60}^+, and C_{70} have also been detected. These detections have been made in a variety of interstellar and circumstellar locations. Molecules with six or more atoms (and including carbon atoms) are conventionally called *complex organic molecules*.

CH CN CH^+ OH CO H_2 SiO CS SO SiS NS C_2 NO HCl NaCl AlCl KCl AlF PN SiC CP NH SiN SO^+ CO^+ HF N_2 CF^+ PO O_2 AlO CN^- OH^+ SH^+ HCl^+ SH TiO ArH^+ NS^+ HeH^+

H_2O HCO^+ HCN OCS HNC H_2S N_2H^+ C_2H SO_2 HCO HNO HCS^+ HOC^+ SiC_2 C_2S C_3 CO_2 CH_2 C_2O MgNC NH_2 NaCN N_2O MgCN H_3^+ SiCN AlNC SiNC HCP CCP AlOH H_2O^+ H_2Cl^+ KCN FeCN HO_2 TiO_2 CCN Si_2C S_2H HCS HSC NCO

NH_3 H_2CO HNCO H_2CS C_2H_2 C_3N HNCS $HOCO^+$ C_3O $l-C_3H$ $HCNH^+$ H_3O^+ C_3S $c-C_3H$ HC_2N H_2CN SiC_3 CH_3 C_3N^- PH_3 HCNO HOCN HSCN HOOH $l-C_3H^+$ HMgNC HCCO CNCN HONO

HC_5N HCOOH CH_2NH NH_2CN H_2CCO C_4H SiH_4 $c-C_3H_2$ CH_2CN C_5 SiC_4 H_2CCC CH_4 HCCNC HNCCC H_2COH^+ C_4H^- CNCHO HNCNH CH_3O NH_3D^+ H_2NCO^+ $NCCNH^+$ CH_3Cl

CH_3OH CH_3CN NH_2CHO CH_3SH C_2H_4 C_5H CH_3NC HC_2CHO C_5S HC_3NH^+ C_5N HC_4H HC_4N $c-H_2C_3O$ CH_2CNH C_5N^- $HNCHCN$ SiH_3CN

CH_3CHO CH_3CCH CH_3NH_2 CH_2CHCN HC_5N C_6H $c-C_2H_4O$ CH_2CHOH C_6H^- CH_3NCO HC_5O

$HCOOCH_3$ CH_3C_3N C_7H CH_3COOH H_2C_6 CH_2OHCHO HC_6H CH_2CHCHO CH_2CHCN NH_2CH_2CN CH_3CHNH CH_3SiH_3

CH_3OCH_3 CH_3CH_2OH CH_3CH_2CN HC_7N CH_3C_4H C_8H CH_3CONH_2 C_8H^- CH_2CHCH_3 CH_3CH_2SH HC_7O

$(CH_3)_2CO$ $HO(CH_2)_2OH$ CH_3CH_2CHO CH_3C_5N CH_3CHCH_2O CH_3OCH_2OH

HC_9N CH_3C_6H CH_3CH_2OCHO CH_3COOCH_3

C_6H_6 C_3H_7CN

$c-C_6H_5CN$

---

allowing for deuterium and other isotope insertion in hydrocarbon species increases the number of species and necessary reactions enormously.

Astrochemists have proposed ingenious schemes of reactions in the gas to account for the presence of many of the two hundred or so species of detected interstellar molecules and for some of the molecules with isotope substitutions, and the validity of these schemes has been tested in many laboratory experiments. The identification of a very rich cosmic chemistry is a triumph of modern astronomical techniques, and we give in Box 6.3 a brief description of how these detections are made.

The evaluation of the chemical routes by which many of these molecules are made in the gas, and evaluating them in the laboratory, has been one of the major intellectual achievements in astronomy over the last half century, involving laboratory and theoretical chemists, and observational and theoretical astronomers. These schemes have been tested in large computer calculations that follow the chemistry as it occurs in networks of thousands of reactions. Many of these ideas have been successful in estimating the abundances of molecular species that are in reasonable agreement with abundances obtained from astronomical observations.

However, some schemes fail to survive close inspection. This may happen simply because we have an imperfect knowledge of chemistry, so that the proposed reaction doesn't work as had been assumed, or the proposed reaction scheme is valid but simply doesn't make enough of the desired product molecular species. However, it may be that no gas phase pathway can be found to a particular detected molecular species. If we are sure that no such pathway exists or that the pathway is inadequate then we have to consider alternatives to chemistry in the interstellar gas. The most plausible alternatives are that reactions occur on the surfaces of dust grains and in the ices that may accumulate on those grains in dark clouds. We'll discuss these ideas in Chapters 7 and 8. However, the calculations based on reactions in the interstellar gas have been remarkably successful, confirming that reactions in the interstellar gas are responsible for forming many of the simpler interstellar species.

6.5 HOW IMPORTANT IS DUST IN MAKING MOLECULES IN INTERSTELLAR GAS?

In this chapter, we have concentrated mainly on the chemistry that occurs in two types of interstellar cloud: diffuse clouds and dark clouds. The important difference between them is the density of the gas, and – since the gas and dust grains are mixed

---

Box 6.3   Detecting molecules in space

To date, over 200 molecular species involving 16 different elements have been detected in space, with the species ranging in size from 2 to 70 atoms (see Table 6.3). About one third of these molecular species have also been detected in external galaxies, with the number of atoms per molecule up to 12. The molecular inventory in proto-planetary discs is relatively sparse, adding up to 36 species (including isotopologues, *i.e.*, molecules that differ only in their isotopic composition), the most "complex" molecules being methanol (CH_3OH), and methyl cyanide (CH_3CN). Only a handful of molecules has been detected in the atmospheres of exoplanets (*i.e.*, planets outside the solar system). The vast majority of all these detections have been made by means of radio-astronomical techniques.

Every molecule in the Universe makes its own characteristic motions based on the disposition of protons, neutrons, and electrons within it. These motions give rise to electromagnetic waves observed by radio telescopes such as the IRAM 30 m telescope. This telescope has been the most successful in making new detections of astronomical molecules. (It is a facility of the Institut de Radioastronomie Millimétrique, a collaboration between France, Germany, and Spain.)

When the signals are displayed on a screen they form a *spectrum*, that is, the representation of the radiation broken down into its constituent wavelengths. This spectrum reveals the chemical makeup of the astronomical region observed, through a sequence of peaks and valleys, each wiggle in the spectrum proving that some molecule emitted radiation of a particular wavelength.

Radio telescopes are built in all shapes and sizes based on the kind of radio waves they receive. The most versatile and powerful type of radio telescope has a large parabolic dish, a shape that forces incoming radio waves to bounce up to a single point above it, called a *focus*. These radio waves travel across the galaxy and through the Earth's atmosphere relatively undisturbed. When large telescopes probe the Universe, the faint radiation they gather may have come from objects millions or billions of light years away. Because radio waves carry little energy, and cosmic radio sources are extremely weak, radio telescopes are among the largest telescopes. The Green Bank Telescope in Virginia (USA) is the largest fully steerable single dish telescope, 100 m across. Unfortunately, being so huge, these antennae easily collect radio interference from electronic devices (such as mobile phone signals). For this reason, radio telescopes are often built in valleys or on mountaintops far from cities, using the surrounding mountains and trees as a shield against radio frequency interference.

Another important reason for radio telescopes to be huge is angular resolution, the ability to distinguish fine details in the sky. To have its resolution similar to optical telescopes, a radio telescope antenna needs to be much larger than an optical telescope. To avoid the construction of impractical devices, the signals received by a group of telescopes spread over a large area can be combined, making them operate together as one gigantic telescope.

(*continued*)

Box 6.3 *(continued)*

This technique is called *interferometry*. Two telescopes at a known distance apart will receive incoming radio waves at slightly different times (this is called *phase shift*). When the signals are combined, they will not overlap, and being shifted, give rise to interference fringes, a succession of bright and dark spots (called an *interferogram*), whose relative locations depend on the arrival time shifts of the waves. The bright regions are places where the two waves add together, the dark areas where they are subtracted from one another. Measuring the interferogram accurately, we can determine the location of the signal in the sky with very high precision, thus obtaining a high-resolution image.

Operating together, the two telescopes can observe only a single point in the sky. Fortunately, the Earth rotates with respect to the sky, and during its rotation the relative positions of the telescopes shift with respect to the astronomical signal, observing a slightly different location. So as the time goes by, an initial discrete set of image points turns into a more solid image. Increasing the number of telescopes, we get many points, one for each pair of telescopes. Moreover, the farther the telescopes are apart, the greater phase shift they see, resulting in a finer detail view of the sky. This is the reason why modern radio instruments are constructed as well-spaced arrays of antennae.

The Atacama Large Millimeter Array (ALMA) is the world's most powerful observatory for studying the Universe at the long-wavelength millimetre and sub-millimetre spectral ranges. It is designed to observe the most distant and most ancient objects ever seen, and to look deeply in star-forming regions and in proto-planetary environments. ALMA has 66 antennae, fifty-four 12 m diameter antennae and twelve 7 m diameter antennae (Figure 6.3). These antennae can be moved with suitable transporter trucks, and are able to be re-positioned to adapt the intertelescope distances to the required observational needs.

Observations with the ALMA have proven to be exceptional tools for the detection of new molecular species in the interstellar medium and for the study of their chemical history and interaction with their physical environment. In Figure 6.4 about 700 emission lines of molecules energized by starlight in the Cat's Paw Nebula are shown. These molecules include methanol, ethanol, methylamine, and glycolaldehyde. The Cat's Paw is located about 4300 light years away, and is a region in which massive stars are currently forming.

In icy grains, molecules are stuck in position and cannot rotate. However, they can still vibrate. Molecular signatures take the form of infrared spectra, which rely on the vibrations of the embedded molecular species energized by passing starlight. These spectra are collected by infrared telescopes on board orbiting satellites – such as the Spitzer Space Telescope (launched in 2003 and still in operation in 2019) or the James Webb Space Telescope to be launched in 2021. Such telescopes need to be located in space otherwise the water vapour in the atmosphere will absorb a large fraction of the infrared radiation. The infrared spectral window hosts a very large number of the most important atomic and molecular transitions.

Figure 6.3 The ALMA telescope. The Atacama Large Millimetre Array (ALMA) is an array of 66 radio telescopes operating at millimetre and sub-millimetre wavelengths, and is located in the Atacama Desert, Chile, at an altitude of 5000 m. This site is above most of the water vapour in the atmosphere, and is radio-quiet. The telescopes are linked and can act together as if they are a single telescope of size equal to that of the array. Consequently, ALMA has unprecedented angular resolution and sensitivity, and is an important facility for the study of star and planet formation (credit: ESO/Y. Beletsky).

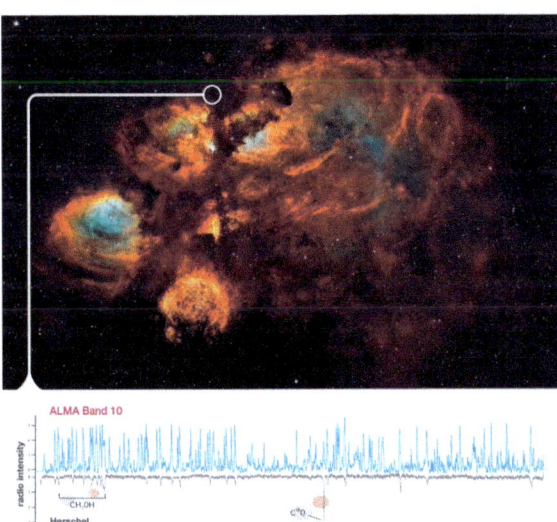

Figure 6.4 Emission lines from the Cat's Paw Nebula. The image shows the Cat's Paw nebula and a very small part of the sub-millimetre spectrum, taken by ALMA. The wealth of information in this spectrum shows that this region is very rich in complex molecules (credit: S. Lipinski/NASA & ESA, NAOJ, NRAO/AUI/NSF, B. McGuire *et al.*).

together – this means that there is a difference in the amount of dust. Diffuse clouds are of relatively low density, and have relatively low amounts of dust. So the extinction of starlight is relatively low in diffuse clouds, and so starlight can penetrate these clouds. Dark clouds are dense enough that radiation is severely extinguished inside them, and in particular ultraviolet starlight is almost completely excluded.

Both types of cloud develop a rich variety of molecular species. We understand fairly well how most of these species arise from the chemistry. We can see in Tables 6.2 and 6.3 a few of the species that are detected in these two types of cloud.

However, there is an important difference between these sets of molecules. Diffuse clouds show mainly molecules containing two and three atoms. The molecular species H_2CO, CH_3OH, C_3H_2, and CH_3CN have also been detected in diffuse clouds.

By contrast, molecular species in dark clouds can have much greater complexity. Some species detected in dark clouds have as many as twelve atoms. Many of these larger species cannot be made efficiently by reactions in the interstellar gas, and they are probably formed by reactions in the ices on the surfaces of dust grains in dark clouds. We'll discuss the chemistry of interstellar ices in Chapter 8.

Our main conclusion, therefore, about the roles of dust grains in the interstellar medium is that they have an important (if passive) role in increasing the chemical complexity of interstellar molecules. They do this by excluding starlight, especially ultraviolet starlight, from the interiors of clouds, rendering them dark. Some of these larger species seem to be related to molecules of biological interest. We'll return to that topic in Chapter 10.

CHAPTER 7

Catalytic Chemistry in Space? Reactions on Bare Dust Grains

Catalysts are substances that can increase the rate of a chemical reaction enormously, and yet are not consumed in the process. At present, for example, our modern society depends almost entirely on catalysts to enhance the rates of production in various processes in chemical engineering, in particular in the production of fuel for internal combustion engines. In the oil industry, the long hydrocarbon molecules that make up crude oil have a high boiling point and are unsuitable as fuel. But these long molecules can be converted using a process known as *fluid catalytic cracking* into the shorter molecules that make up appropriate fuels for our cars and lorries. The active catalyst in this process is *zeolite*, a crystalline material containing aluminium and silicon that has some structural similarity to silicates.

Could catalysts be operating in interstellar gas clouds? Atoms and molecules are certainly present in the interstellar gas, as we have seen in the previous chapter, and there are dust grains mixed in the gas. We know something about the physical and chemical properties of this dust, as we've described in Chapters

Dust in Galaxies
By David A. Williams and Cesare Cecchi-Pestellini
© David A. Williams and Cesare Cecchi-Pestellini 2020
Published by the Royal Society of Chemistry, www.rsc.org

2 and 3. Is it possible that these dust grains could have a role to play in catalyzing some chemical reactions in the interstellar gas? Could interactions between gas atoms or molecules and dust grains enhance the abundances of some interstellar species?

These questions have been around for about seventy years. The idea of surface reactions occurring on the surfaces of interstellar dust grains was first put forward because it was realised that reactions in the interstellar gas simply aren't capable of producing the amounts of molecular hydrogen observed to be present in the diffuse clouds of the Milky Way galaxy. Molecular hydrogen can certainly be made in reactions involving species in the gas, but not in the amounts detected in Milky Way clouds. Could reactions of atomic hydrogen on dust grains could be a better source of molecular hydrogen? Very reluctantly, astronomers began to accept the idea that molecular hydrogen might be formed somehow on the surfaces of dust grains, although there were initially no reliable experiments or fundamental theoretical calculations to support this conclusion.

As we've seen in the preceding chapter, molecular hydrogen is absolutely fundamental to astrochemistry. The main entries to the chemical networks, forming OH and H_2O from O^+, and CH, CH_2, and CH_3 from C^+, all depend on reactions with H_2. Molecular hydrogen can be regarded as the seed molecule for all of interstellar chemistry, and so we might very properly call molecular hydrogen the *seminal molecule for astrochemistry*. Therefore, it was difficult for astronomers to accept that the exquisite achievement of defining the astrochemical networks involving thousands of reactions in the interstellar gas, based on accurate and hard-won experimental data, actually rested on an almost complete lack of reliable information about the formation of hydrogen molecules in interstellar clouds. As a result, while the idea of catalysis of molecular hydrogen on dust grains was – in the absence of any supporting data – grudgingly accepted, the formation of any other molecules in this unorthodox way was discouraged. In fact, as we'll see in this chapter and in those that follow, the surfaces of interstellar dust grains are chemically important in various ways.

Fortunately, definitive answers to the question of hydrogen formation on surfaces have become available during the last two decades, as both laboratory and computational techniques have improved. It now seems clear that molecular hydrogen can be made on the surfaces of interstellar dust grains made of silicates and carbons, and that the rates of formation are probably large enough to account for the observed abundances in diffuse and dark clouds. Other molecular species detected in interstellar clouds may also be formed in surface reactions on dust grains.

The interaction of atoms and molecules from the gas with dust grains has also turned out to be an important topic that helps us to understand how mantles of ice on dust grains may grow in dark clouds. The chemistry of these ices has become very important in the study of complex molecules and their connection with astrobiology. These are topics that we'll discuss in Chapters 8 and 10.

7.1   FORMING AND DESTROYING THE SEMINAL MOLECULE

Rather perversely in a section devoted to molecular hydrogen formation, we start by considering how this molecule is destroyed in diffuse interstellar clouds. But the idea is simple: if we know how quickly the molecule is being destroyed, then we can assess whether a formation process can be fast enough to provide the amount of molecular hydrogen that is detected in these clouds. The question is this: are the rates of formation by any proposed processes adequate to account for the observed abundance of H_2 in diffuse clouds?

It's a bit like filling a bath with water to a desired level. Suppose that the tap supplies water at a certain rate and an unplugged drain allows the water to flow away. If we know the rate of outflow though the drain, then we'll also know whether the inflow is either too fast (in which case the bath will overflow) or too slow (in which case the water level will never rise to the desired level).

Do we know how fast hydrogen molecules are destroyed in interstellar clouds? Yes, we do. The destruction of hydrogen molecules by ultraviolet starlight is closely related to the way they are detected. We know that hydrogen molecules are present in interstellar diffuse clouds because they absorb ultraviolet starlight in a characteristic manner that is a unique signature of molecular hydrogen. The effect of the energy absorbed from starlight by the molecule is to raise the molecule from its lowest energy state, the ground state, to an excited state. Then what happens? The molecule wants to drop down again to the ground state, and it does this after being in the excited state for less than one millionth of a second.

However, there is about a twenty percent chance that when the molecule drops down to the ground state, it will simply fall apart; the atoms separate, therefore the molecule doesn't exist anymore. In principle, the rate of destruction of hydrogen molecules in diffuse clouds that are irradiated by ultraviolet starlight is related to the rate at which hydrogen molecules absorb the starlight. It's a known quantity.

These excitation and destruction processes are described in more detail in Box 7.1.

Astronomers tried very hard to find ways of forming molecular hydrogen using only atoms, ions and electrons in the gas. There are several possible methods. The apparently obvious route:

$$H + H \rightarrow H_2$$

in which two hydrogen atoms in the gas collide with each other is calculated to be almost infinitesimally slow and can be ignored. Therefore, a three-body collision (which is the common type of reaction occurring in the Earth's atmosphere) was considered:

$$H + H + H \rightarrow H_2 + H$$

It is certainly a type of reaction that works. The idea is that one of the partners carries away some energy from the other pair, leaving them without enough energy to escape from each other. But for the reaction to work, densities that are enormously higher (at least around one billion times higher) than those found in diffuse interstellar clouds are required. The three-body reaction simply doesn't work at the low densities of diffuse clouds, because the chance of a third atom meeting a colliding pair of atoms is really very small unless the gas density is very high. It may be important in some extremely dense parts of the interstellar medium.

Box 7.1 How are interstellar hydrogen molecules destroyed?

Hydrogen molecules are identified in diffuse interstellar clouds because they absorb ultraviolet starlight. If there is a hot star behind the cloud, then the hydrogen molecules absorb starlight at particular wavelengths, creating a recognisable pattern in the transmitted radiation, which uniquely identifies the absorber as H_2. The absorption spectrum also allows the number of H_2 molecules in the diffuse cloud to be estimated. The arrangement of the hot star, the cloud, and Earth is shown in Figure 7.1 (top), and a typical absorption spectrum for H_2 in a diffuse cloud is given in Figure 7.1 (bottom).

Hydrogen molecules in cold gas move around, and they also vibrate and rotate. In interstellar diffuse clouds they are normally in the ground electronic state, and in the lowest vibrational level (we'll label this lowest vibrational level in the ground electronic state $v'' = 0$), and in one of a few rotational levels (labelled $J'' = 0, 1, 2,$ *etc.*). For convenience, we'll ignore rotation here.

When a hydrogen molecule absorbs ultraviolet light, it jumps from the ground electronic state with $v'' = 0$ up to an excited electronic state, and to a particular vibrational level in the excited state. We'll call the vibrational level in the excited state v', and it can take values $0, 1, 2,$ *etc.* We show in Figure 7.2 an energy level diagram showing the ground electronic state, here called X, and one excited electronic state, here labelled B.

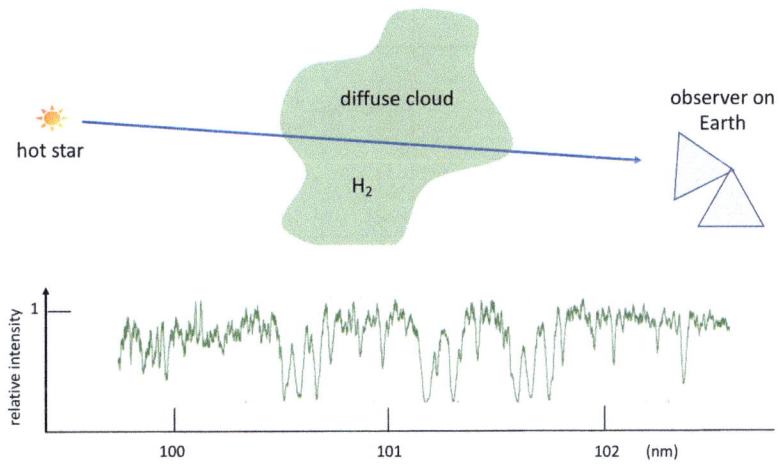

Figure 7.1 (Top) Geometry, and (bottom) absorption spectrum of H_2 in a diffuse cloud. The top image shows schematically how light from a hot star passes through a diffuse interstellar cloud on its way to Earth. The bottom image shows a small part of the actual ultraviolet spectrum of the radiation of the hot star, as it is received at Earth. Most of the radiation has been removed because hydrogen molecules in the diffuse cloud between the star and Earth have absorbed the radiation.

(continued)

Box 7.1   (*continued*)

   The upward arrow in Figure 7.2 represents absorption from the $v'' = 0$ of the ground electronic state, X, into the excited electronic state, B, in one of the vibrational levels, v', of state B. The absorption detected in an astronomical observation corresponds to a transition exactly like this one. In practice, it is slightly more complicated because this single line between two vibrational levels, one in X and one in B, is actually replaced by several lines, because we have ignored the closely spaced rotational levels that are present in all the electronic states.

   The excited molecule has a very short life, and very quickly jumps down to the ground state, X. It ends up in one of the vibrational levels of the ground state. There are 14 separate levels in which the two atoms are bound together in the H_2 molecule; the highest of these is $v'' = 13$. But there is a set of v'' levels above $v'' = 13$ and these are separate, but continuous. These continuous levels have enough energy in them that if the H_2 molecule enters one of these continuous levels it simply falls apart. The downward arrows in Figure 7.2 represent this relaxation process from B to X. One of these downward transitions falls into a v'' level representing the bound molecule. The molecule in level v'' then jumps down in much slower steps until it arrives in the level $v'' = 0$.

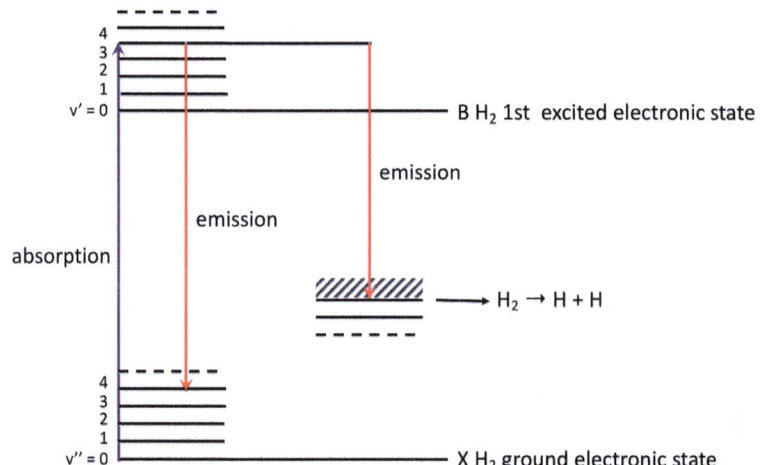

Figure 7.2   The H_2 energy level diagram. The figure shows the ground state, X, and the excited state, B, of molecular hydrogen. Each state has associated with it a number of vibrational levels. The molecule in the interstellar medium is normally in the lowest vibrational level of the ground state, shown here as $v'' = 0$. The molecule can absorb energy from ultraviolet starlight, and jump to excited state B in one of the vibrational levels, labelled here as v'. The molecule then jumps down again to the ground state, but to an excited vibrational level. If this level v'' is less than 14, the molecule survives, but if it falls into a higher level the molecule simply falls apart.

> The other downward transition illustrated in Figure 7.2 represents a molecule falling into the continuous levels of the ground state, X. This molecule falls apart, forming two hydrogen atoms. Detailed calculations show that for unshielded hydrogen molecules in the typical interstellar radiation field, about twenty percent of all the excitations end up destroying the molecule.
>
> We know the intensity of the unshielded radiation field, and we know how this intensity changes as the radiation penetrates the diffuse cloud. We also know how a fraction of these molecules is destroyed each time they are excited. So the rate of destruction of hydrogen molecules in the interstellar radiation field can be calculated.

There are two methods that do work well under the conditions that we find in diffuse clouds. One involves the H^- ion. It may form when an electron collides with and attaches to a hydrogen atom (a rather slow process):

$$H + e \rightarrow H^-$$

This negative ion may then meet a hydrogen atom and form a hydrogen molecule, spitting out the electron. Therefore, the electron is returned to the gas and is available once again; so it is actually a catalyst in the H_2 formation reaction:

$$H^- + H \rightarrow H_2 + e$$

where e represents an electron. There's a similar type of reaction involving hydrogen ions, H^+, rather than electrons:

$$H + H^+ \rightarrow H_2^+$$

followed by:

$$H_2^+ + H \rightarrow H_2 + H^+$$

Here, the hydrogen ion, H^+, is also a very simple catalyst. It helps the reaction to proceed but isn't used up in the process. All the basic chemical data for these reactions has been calculated quite accurately, so the rates of H_2 formation by each of these processes are well known. In the interstellar medium, they are both rather slow, because the initial step in each case – the probability of addition of either an electron or a hydrogen ion to a hydrogen atom – is low. Also, the intermediates H^- and H_2^+ are themselves quite easily destroyed by radiation of long wavelengths (that is, the infrared); there's plenty of such radiation in the interstellar medium from low-mass stars.

The outcome of these studies is that the H^- and H_2^+ schemes, attractive though they are, simply aren't fast enough to provide enough molecular hydrogen in diffuse clouds of the Milky Way galaxy, as revealed by observations. This conclusion encouraged astronomers to consider alternatives, in particular the formation of hydrogen molecules on the surfaces of interstellar dust. Could dust act as a catalyst in the formation of molecular hydrogen? As we'll see, that idea is now supported by both laboratory and theoretical work, and there is no doubt that surface reactions do occur in interstellar clouds, and that they are capable of providing the amount of H_2 observed in diffuse clouds.

Of course, this conclusion doesn't mean that these H^- and H_2^+ schemes are 'wrong'. They are simply inappropriate for the Milky Way and for many other galaxies, but they may be applicable to other types of astronomical region, especially those regions where dust doesn't exist or is under-abundant. The early Universe must have been such a place. Dust couldn't have formed in a Universe before the stars had formed. The early Universe consisted almost entirely of hydrogen atoms and helium atoms and their ions. In such conditions, molecular hydrogen would still have formed through the H^- and H_2^+ routes, and this H_2 must have helped in the formation of the first stars.

7.2 SURFACE REACTIONS ON DUST: HOW THEY WORK

The basic idea is simple: we want one H atom to collide with a dust grain and be captured on the surface, so that another H atom can find the first one and react with it. A process like that should give much more time for the pair of H atoms to be in contact and combine to form H_2 than if they were zooming past each other at high speed in the gas. The dust grain also provides a sink for energy to be taken from the colliding pair, so that they don't have enough energy to separate and are therefore trapped in a molecule. The newly-formed molecule may either be ejected from the surface at once, or reside there until it is ejected by some other process. In the former case, the product molecule may be vibrating and rotating.

The second H atom may either collide directly with the first atom, with little interaction with the surface, or it may become attached to the surface and – if it isn't too strongly bound to the surface – roam over it until it finds the first atom and interacts with it. These ideas for H_2 formation on a surface are summarized schematically in Figure 7.3.

There are some limits on the interaction of the first atom with the surface. If that interaction is too weak then the atom may leave the surface before the second atom arrives, so no molecule could be formed. On the other hand, if the first atom is too strongly tied to the surface, then the second atom may not be able to react and tear the first one away from the surface as part of an H_2 molecule. Obviously, the nature of the surface – its chemical nature, whether it is rough or smooth, whether it is porous or solid – may be critical.

What can be done to explore these ideas? There are two approaches. We can make experiments, or we can make complex calculations. Both approaches have advantages and disadvantages.

Experiments must be carried out in an ultrahigh vacuum, but still can't match very well the conditions in interstellar diffuse clouds. The experiments must be done on relatively large surfaces

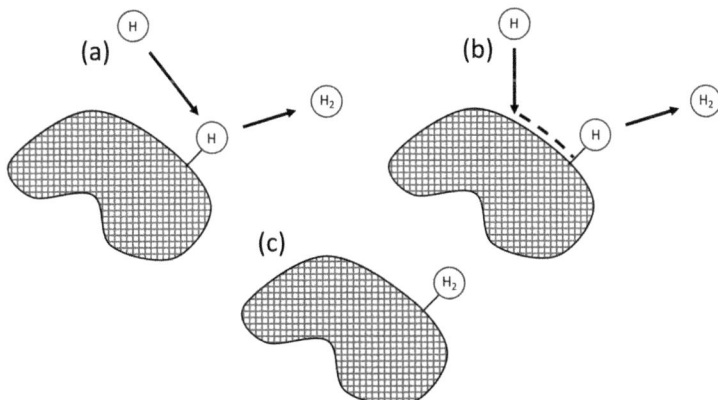

Figure 7.3 Various ways in which H_2 could be formed on the surface of a dust grain. In (a), an H atom from the gas collides directly with an H atom that is attached to the grain and forms an H_2 molecule, which is ejected off the surface. In (b), the H atom from the gas collides with the grain surface, and moves over the surface until it finds the first H atom, then forms a molecule, which is ejected. In (c), the molecule that is formed in either case (a) or (b) isn't ejected but remains bound to the surface.

rather than tiny dust grains. These surfaces probably contain a wide variety of structures, so we don't really know what the surface is really like in detail. On the other hand, experiments can use a variety of materials to represent the surfaces of dust grains, so an experimenter can test a variety of types of dust grain materials. Also, experimental results are always convincing – at least, to theoreticians!

Theoretical approaches are necessarily highly idealized and simplified, and cannot represent the actual surfaces of real dust grains in the interstellar medium. Ideally, theoreticians use a well-defined structure, such as graphene, as the surface. It's more difficult to introduce imperfections in the surface. However, the conditions assumed in the calculations can be made like those in the interstellar clouds. We do gain understanding from theory, too, because the calculations can be repeated as often as required to understand the sensitivity of the results to changes in, say, temperature or gas density. More details are given in Box 7.2.

Over the last two decades, techniques in the laboratory and in theoretical chemistry have improved enormously, and many useful experiments and theoretical calculations have been made to help us understand surface chemistry on dust grains. What have we learned?

Experiments and theory both agree that H_2 formation can occur on the surfaces of dust grains. The more realistic the theoretical model, the more likely the formation is to occur. Some experiments find that the newly-formed H_2 molecules reside on the surface and can be removed after formation by heating. Other experiments show that the molecules are formed in vibrationally excited states and are ejected from the surfaces with high velocity. So it seems that the way the experiment is designed allows us to examine different mechanisms and products arising in the chemistry. There are some differences between reactions on carbons and silicates – the likely materials of dust grains – but both seem able to catalyze the formation of H_2. Some experiments show that H_2 formation can continue even when the temperature of the dust grains is raised considerably above the value expected for dust grains in diffuse interstellar clouds.

Neither the experiments nor the theory can really describe reactions on the surfaces of actual dust grains in the

Box 7.2   Experiments and theory

Experiments

There are now many experiments in different laboratories that demonstrate conclusively that molecular hydrogen can form in reactions of atomic hydrogen on surfaces. In these experiments, the surfaces are chosen to be representative of interstellar dust grains, usually carbons or silicates of some kind. Of course, these surfaces are enormous compared to those of very tiny particles of interstellar dust.

The experiments are simple in concept, but rather complicated in details, so much so that it is only about twenty years since the first successful demonstration of H_2 formation. What are these complications?

The gas in interstellar clouds has very low density and very low temperature. So the experiments have to try to match those conditions as far as possible, and these are severe technical demands on the experimenters. These demands require the experiments to be conducted at ultrahigh vacuum and with the solid surface at a very low temperature.

Then the atomic hydrogen needs to be directed onto the cold surface, and a method of detecting and measuring the number of hydrogen molecules formed at the surfaces developed. In the laboratory, hydrogen is available as molecular hydrogen, so a way has to be found of turning molecular hydrogen into atomic hydrogen, and then forming a beam of atomic hydrogen directed at the cold surface. The method usually used is to use radio waves of a certain frequency to shake hydrogen molecules apart, generating hydrogen atoms. This process isn't totally efficient, so the beam always contains H_2 molecules as well as H atoms.

How can we tell if H_2 has actually formed in reactions at the surface or is simply H_2 carried in the beam? There are two ways that have been used. In the first, two separate beams of mainly atomic hydrogen were directed at the surface. However, these were of different isotopic varieties. One beam was of H and H_2, while the other beam was a deuterium beam of D and D_2. The detection of HD would imply that a reaction had taken place between H from one beam and D from the other beam, since HD existed in neither beam.

The other method that has been used detect newly formed molecular hydrogen is to assume that the newly formed H_2 would be in a highly excited vibrational state. Spectroscopic methods can then help to distinguish this newly formed H_2 from H_2 that existed in the beam. In this type of experiment, only one hydrogen beam is required.

The final necessary component of these experiments is to have some way of counting the number of newly formed hydrogen molecules (HD in the first type of experiment, and H_2 in the second). The usual method is to detect only products of the correct mass, a mass of 3 atomic units for HD (since D is twice as massive as H) or 2 atomic units for H_2. This is done by ionizing the newly formed molecules and using a combination of electric and magnetic fields to distinguish between the various masses.

The first experiment to demonstrate molecular hydrogen formation was carried out in 1997 by Valerio Pironello and Gianfranco Vidali and colleagues at Syracuse University, New York. They used two atomic beams,

(continued)

Box 7.2   (*continued*)

one of hydrogen and one of deuterium, aimed at the cold surface (initially a silicate). They assumed that newly-formed HD molecules would be weakly bound to the surface and retained there, so they warmed the surface very gently and detected the HD emerging during the warming.

An alternative view is that the formation process is really quite energetic, and that the newly formed molecules would be ejected promptly from the surface and would be vibrationally excited. A separate experiment by Stephen Price and colleagues at University College London in 2002 showed that H_2 formed on graphite was indeed highly vibrationally excited, and was ejected from the surface with high velocity.

These experiments showed that both types of molecular hydrogen formation occur. In one case the newly formed molecule is retained on the surface and released as the surface is warmed after the formation process is completed, while in the other case the newly formed molecules are released during the reaction on the surface. An experiment conducted by Jean-Louis Lemaire and colleagues at the Observatoire de Paris and the Université de Cergy Pontoise in 2010 showed that both types of reaction can occur at once. In their experiments on molecule formation on a silicate (olivine) surface, they found that the prompt and energetic release of newly formed molecules occurred for a wide range of surface temperatures, while the delayed release occurred over a rather narrower range.

Evidently, all these experiments tell us that molecular hydrogen formation at low temperatures on cold surfaces can occur efficiently. This is a valuable result, but doesn't tell us much about how the process depends in detail on the nature of the surface, at the atomic scale. This is where a theoretical approach can be helpful.

Theoretical Methods

According to the ideas expressed in Figure 7.3, there are various stages to the formation of molecular hydrogen on the surface of a dust grain. First, an H atom must stick to the surface. A second H atom may approach the first one directly, and a reaction may take place, or the second may also stick somewhere on the surface and then be mobile enough to find the first atom and then react with it. Then the newly formed molecule may be retained on the surface or ejected into space. From a theoretical point of view, the various stages in molecule formation on a surface are (i) sticking, (ii) mobility, (iii) reaction, and (iv) ejection. All of these processes have to be considered at the atomic level.

Fortunately, these problems can now be addressed using the appropriate mathematical language to describe atomic behaviour. This language is, of course, quantum mechanics. It's no good simply to consider atoms as balls colliding with each other in ways that seem familiar to us. Atoms don't necessarily behave like that. They have their own rules to obey, and these rules – which can be quite weird – can be found using quantum mechanics. With the use of fast computers, it is now possible to make really accurate calculations of the way that atoms react with each other and with surfaces. To make calculations of fairly realistic situations, however, requires rather lengthy calculations. In principle, all the processes (i)–(iv) can be evaluated.

However, the more realistic the situation the harder the calculation. For example, is the surface a regular crystal, or is it highly disordered? Are other atoms inserted into the lattice?

Most of the work has been done assuming that the surface is like that of graphene. The basic calculation suggests that H atoms cannot bond chemically with the graphene surface. However, if the graphene surface is imperfect in some way, then H atoms can bond easily and would then be available for chemical reactions as indicated in Figure 7.3.

Detailed studies of the H_2 formation reaction on a graphene surface indicate that the reaction should occur almost every time two H atoms meet on the surface, and that the newly-formed H_2 molecule will be ejected from the surface at high speed and with a large amount of vibrational energy. Similar conclusions can be reached for H_2 formation on silicate surfaces although those surfaces are more complicated; it may be necessary for several H atoms to be adsorbed on the surface and involved in the H_2 formation.

Real interstellar grains aren't expected to have perfect crystal formation. Indeed, as we have seen in Chapter 2 the grains are likely to be amorphous, possibly with small regions of crystallinity within them. Grains like that will contain many steps, kinks, and other defects, such as nearby adsorbed atoms. These will all have their own chemical behaviour.

interstellar medium. Chapters 2 and 3 tell us that the material of dust grains is usually amorphous rather than crystalline, but we don't really know whether the dust grains in interstellar clouds are porous or uniformly solid, rough or smooth, and we don't really know what impact these qualities may have on the formation efficiency. What we need to do is to compare what we have learned from experiments and theory with the results of observations. We need the overall H_2 formation efficiency predicted by these studies. We can work this out for a given distribution of sizes of dust grains, assuming that all grains, large and small have the same reaction efficiency.

The result we find is this: the H_2 formation rate we calculate for a population of dust grains ranging from small to large with many more small grains than large can match the destruction rate we infer from the observations, but for this to happen *we need almost every H atom that arrives at the surface of a dust grain to be converted to an H_2 molecule.*

This, then, is our final result. The formation of molecular hydrogen in reactions on dust grain surfaces in interstellar diffuse clouds needs to be an efficient process. Both experiment and theory suggest that it is.

7.3   CAN MOLECULES OTHER THAN H_2 FORM ON DUST GRAINS?

In the previous section we have shown that experiments and theory support the idea that it is possible to add one hydrogen atom to another hydrogen atom that is bound to a surface. If we assume that this happens efficiently on the surfaces of dust grains of a wide range of sizes in diffuse interstellar clouds, then the formation rate of hydrogen molecules is fast enough to match the destruction of H_2 by the interstellar ultraviolet radiation field. We might then ask: are there other reactions that take place on the surfaces of interstellar dust grains? It would seem possible that – if we can add one hydrogen to another on a surface to form H_2 – atoms of oxygen, carbon, and nitrogen colliding with dust grains might be converted to hydrides such as OH, H_2O, CH, CH_2, CH_3, CH_4, NH, NH_2, and NH_3, given that hydrogen is so much more abundant (more than a thousand times) than oxygen, carbon, or nitrogen.

Nitrogen hydrides are a case of particular interest. All the nitrogen hydrides, NH, NH_2, and NH_3, have been detected in some diffuse clouds, but the networks of reactions in the interstellar gas describing the chemistry of diffuse clouds (discussed in Chapter 6) fail to produce enough of the nitrogen hydrides to match the observations. On the other hand, these gas chemistry networks seem to work well for OH and CH. The problem is that there is no simple entry route to nitrogen chemistry in diffuse clouds similar to those entry routes for oxygen and carbon.

In fact, there have been a number of laboratory experiments studying the formation of nitrogen hydrides in surface reactions, and these experiments suggest that the successive additions of hydrogen to nitrogen as far as ammonia (NH_3) are all efficient, that is, they occur nearly always when the reaction partners meet on the surface. The successive stages of hydrogenation are:

$$N \rightarrow NH \rightarrow NH_2 \rightarrow NH_3$$

(nitrogen can't hold any more hydrogen atoms than three) but we don't know whether all the product molecules at each stage remain on the surface and can react to form the next stage. If they do remain on the surface until NH_3 is formed and ejected into the gas, then the NH and NH_2 molecules will be formed by the action of the radiation field on NH_3 in the gas:

$$NH_3 \rightarrow NH_2 \rightarrow NH \rightarrow N$$

We can explore these possibilities in an extended model of a diffuse cloud in which nitrogen atoms arrive at the surfaces of interstellar dust grains and react with hydrogen with the same efficiency as in the H_2 formation reaction. We can compare the results with observations of interstellar diffuse clouds in which all of NH, NH_2, and NH_3 have been detected. The results of these studies tell us that at each stage of hydrogenation some of the intermediate product molecules must be ejected into the gas and some retained on the surface so that the next stage of hydrogenation can take place. When the final stage, NH_3, is reached, all those ammonia molecules are ejected from the surface into the gas.

These theoretical studies support the suggestion that the addition of hydrogen atoms to nitrogen atoms does occur at the surfaces of dust grains in diffuse clouds. The observed abundances of NH, NH_2 and NH_3 in the gas of diffuse clouds can be accounted for by such reactions and the ejection of some of the intermediate products into the gas. It seems likely that similar surface chemistry occurs with oxygen and carbon. However, the chemistry of oxygen and carbon atoms in the gas of diffuse interstellar clouds that we described in Chapter 6 is so efficient that it the surface chemistry makes a minor contribution.

7.4 THE GAS–GRAIN INTERACTION IN INTERSTELLAR CLOUDS: INTERSTELLAR ICES BEGIN TO GROW

The detection of water ice mantles on dust grains in interstellar dark clouds was made almost half a century ago. The detected broad absorption feature centred at a wavelength of 3 μm is attributed to O–H vibrations in the H_2O molecule in the ice. Rotational structure that would normally be present for H_2O molecules free to rotate in the gas was absent. Hence, the molecules must be fixed in the ice and not able to rotate.

This discovery has expanded our knowledge of the interstellar medium, transformed astrochemistry by extending molecular complexity, and even opened up the possibility of biology having some of its origins in interstellar space. We'll discuss all those topics in Chapters 8 and 10. In this chapter we have already discussed the significance of surface reactions on the chemistry of diffuse clouds and we have seen that observations indicate that the retention of some species on the surfaces of dust grains is occurring. In other words, we are seeing the initial stages of the build-up of ices on dust grain surfaces. This happens naturally in diffuse clouds and develops as we follow the transformation of diffuse clouds into dark clouds.

When the discovery of interstellar water ice was made, it became immediately clear that this ice wasn't simply deposited from water molecules in the gas. Yes, water molecules are present in the gas in dark clouds but at such a low abundance that icy mantles containing a large amount of available oxygen could not be frozen out on grain surfaces in a reasonable length of time. The only way to deposit ice mantles in the time available is to form water molecules directly by surface reactions of oxygen atoms with hydrogen. Such reactions might be summarized in this way:

$$O + \text{dust grain} + 2H \rightarrow H_2O + \text{dust grain}$$

Therefore, the detection of interstellar water ice (which occurred in 1973, just six years after the detection of interstellar molecular hydrogen in the gas) is direct confirmation that at least two types of surface reactions were occurring on interstellar dust grains: H_2 formation and H_2O formation. In the case of hydrogen, the molecule is detected in the gas, while for oxygen the H_2O is retained on the grain surface. In diffuse clouds, the water molecule will soon be removed from the grain by the interstellar radiation field, but as the gas density and interstellar extinction increase, the molecule will be retained for longer and eventually the ice mantle will build up. The quantity of ice observed to be on dust grains in interstellar clouds, as measured by the strength of the absorption feature at 3 μm, increases strongly with the amount of extinction. There is a lot of scatter in the observational data for low-extinction clouds, as to be expected, but strong ice absorption at 3 μm is detected in very dense dark clouds.

The chemical composition of interstellar ices will be discussed in detail in Chapter 8, but it's worth noting now that as well as water (H_2O), (the predominant constituent), ammonia (NH_3) and methane (CH_4) are detected in the ice at about a few percent level relative to H_2O. These molecules are also evidence for surface reactions on dust. Other molecules include carbon monoxide (CO), carbon dioxide (CO_2), and methanol (CH_3OH). It seems likely that CO_2 is formed by the reaction of CO with OH (since OH is available from the photodissociation of water):

$$CO + OH \rightarrow CO_2 + H$$

and methanol by the successive hydrogenation of CO:

$$CO \rightarrow HCO \rightarrow H_2CO \rightarrow CH_3O \rightarrow CH_3OH$$

This conclusion is supported by experimental data.

We conclude that the surfaces of dust grains are chemically active. Dust grains *are* effective catalysts in the interstellar medium.

CHAPTER 8

Chemistry in the Freezer: Making Complex Molecules From Simple Interstellar Ices

8.1 SIMPLE ICES IN INTERSTELLAR DARK CLOUDS

We saw in Chapter 6, Section 6.4, that the water molecule – H_2O – can be formed in chemistry that takes place in the gas of interstellar clouds. In fact, water molecules in the interstellar gas were detected very early in the history of astrochemistry. In 1969, Charles Townes and colleagues detected strong water emission in a line with a wavelength of 1.38 cm, in the radio part of the spectrum. This emission arises from freely rotating water molecules in the gas.

Interstellar water *ice*, as distinct from gaseous water molecules, was first identified by Fred Gillett and William Forrest in 1973, and is now widely detected in interstellar dark clouds. Water molecules in ice are fixed in place in the solid material and aren't free to rotate as they are in the gas, so the detection isn't made in the rotational spectrum of water. However, the molecules are able to

Dust in Galaxies
By David A. Williams and Cesare Cecchi-Pestellini
© David A. Williams and Cesare Cecchi-Pestellini 2020
Published by the Royal Society of Chemistry, www.rsc.org

vibrate and can absorb energy at a wavelength of about 3 μm in the infrared part of the spectrum. We saw in Chapter 7, Section 7.4, that this water must be created at the surfaces of dust grains and retained there, forming ice, as there simply aren't enough water molecules in the gas for them to stick to the surfaces of dust grains and form coatings of ice within the typical age of a dark cloud.

Darker clouds have stronger absorption at 3 μm, and therefore a greater thickness of ice. Dust grains in diffuse clouds don't show any signature of water ice in their spectra, so they cannot be made of ice, and the evidence is that they are mainly composed – as we have seen in Chapter 2 – of silicates and carbons.

Why are dust grains in diffuse clouds free of ice while grains in darker clouds are coated with ice? There are several processes that tend to remove water molecules from interstellar ice. Firstly, the temperatures of dust grains in interstellar clouds are very low, about ten degrees above absolute zero of temperature. At these temperatures, evaporation of the ices directly into the gas is entirely suppressed, although ices are readily evaporated if dust grains are heated to about one hundred degrees above absolute zero. Such temperature excursions are very rare. Secondly, starlight may play a role: the absorption of starlight by an H_2O molecule can break the molecule into the separate fragments H and OH, and give each fragment some excess energy. This energy may be enough to allow the fragments to escape from the solid ice, and they may leave the ice in about one percent of all such events, effectively removing one H_2O molecule from the ice. Thirdly, cosmic rays (fast particles pervading the Universe causing ionization in the gas, see Chapter 5, Section 5.3) also pass through dust grains and give up some energy when they do so, heating the grains and therefore possibly causing some evaporation of ices. However, the evaporation is only complete for the smaller dust grains.

In fact, while all these mechanisms may play a role, it seems likely that the transition from "no ice" to "ice" may arise simply because a single H_2O molecule on a dust grain surface experiences a very different environment to that of an H_2O molecule that is part of a solid ice coating. In the former case, the lone H_2O molecule can bind only with the surface of the dust grain. In the latter case, the two H atoms in the H_2O molecule make a special bond (a so-called *hydrogen bond*) with O atoms of neighbouring water molecules in the ice. The result is that each H_2O molecule in the ice is bound by four bonds arranged tetrahedrally, as shown in Figure 8.1.

The time it takes to build up an accumulation of ice on a dust grain, the *freeze-out time*, depends on the density of the gas: the higher the density, the shorter the freeze-out time. In diffuse clouds the density of the gas is low, so the freeze-out time is long. In fact, it is so long that it is longer than the age of the diffuse cloud, which will have dispersed. So there isn't enough time for ice to accumulate on the dust grains in diffuse clouds. On the other hand, the gas density in denser and darker clouds is high, so the ice accumulation time is short, and ice coatings can build up and any mechanisms that tend to remove water molecules from the dust grains are simply overwhelmed in denser regions. Hence, the densest and darkest clouds tend to have the greatest accumulations of ice on their dust grains, and the absorption at the wavelength of 3 μm is strongest for these clouds.

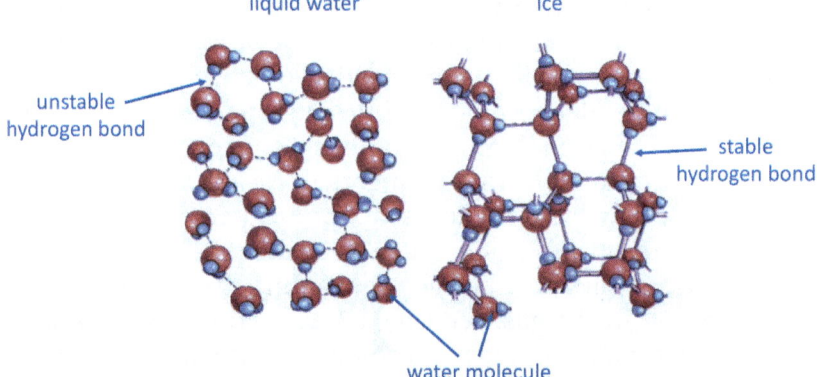

Figure 8.1   Hydrogen bonding in ice. The diagram illustrates the difference between liquid water and solid water ice in terms of the hydrogen bonds between water molecules, each molecule being shown as one oxygen atom (red) associated with two hydrogen atoms (blue). Hydrogen bonds are established between a hydrogen atom in one molecule and the oxygen atom of another molecule. In liquid water, these bonds are relatively weak compared to energies of molecular motion, so molecules in the liquid are always in motion. In solid ice they are relatively strong compared to molecular motions, so the molecules are fixed in place by the hydrogen bonds.

Interstellar ices are very far from being clean. In fact, they are dirty, and probably highly poisonous. Interstellar travellers should not refresh themselves with this stuff! Using methods similar to those by which water ice was detected, other species have been identified, as we saw in Figure 2.6. These infrared spectra all show, as for water ice, pure vibrational spectra, with no rotational components. This is how we know that these materials are ices, and not in the gas.

These spectra show that a variety of simple chemical species is present in interstellar ices. The data in Figure 2.6 also emphasize that the chemical composition of the ice isn't the same on all lines of sight. This seems to imply that it is possible to change the chemical composition of these ices, suggesting that it may be possible to achieve greater chemical complexity by 'processing' these rather simple ices in some way. Detected species in the ice and their typical abundances relative to water molecules in the ice are shown towards low-mass stars and high-mass stars in Table 8.1. We might expect some differences because high-mass stars may provide a more intense radiation field affecting nearby dark clouds, and this stronger radiation may help to convert simple chemicals into more complex molecules.

We can also obtain some information about the structure of interstellar ices: are they chemically uniform or is a particular molecule located at some depth in the ice layer? Let's consider one particular species: carbon monoxide (CO). Table 8.1 shows that CO is an important constituent of the ice mantles. It is detected in interstellar clouds through an absorption band at a

Table 8.1 Typical relative abundances of ices in clouds towards low-mass and high-mass stars.

Component	Low-mass stars	High-mass stars
H_2O (water)	100	100
CO (carbon monoxide)	29	13
CO_2 (carbon dioxide)	29	13
CH_3OH (methanol)	3	4
NH_3 (ammonia)	5	5
CH_4 (methane)	5	2
XCN (cyanides; OCN^-)	0.3	0.6
HCOOH (formic acid)	1–3	1–3

wavelength of 4.7 μm. The onset of CO ice occurs, however, in darker clouds than those in which the onset of water ice occurs. The interpretation that is inferred from this information is that water ice deposition is mostly completed before the CO ice is deposited on the dust grains. Therefore, interstellar ices appear to have a *layered structure* surrounding the dust grain cores: an inner water-rich layer forms early in the evolution of a cloud by hydrogenation of atomic oxygen, frozen at the grain surface, and solid methane (CH_4) and ammonia (NH_3) are also likely to be formed in surface reactions at this stage, through hydrogenation of carbon and nitrogen. During the collapse of the cloud to form a star (see Chapter 9) the increase in gas density in the cloud is so steep that the freeze-out timescale shortens dramatically, inducing a catastrophic removal of the gas, whose main component (after the volatile molecular hydrogen, H_2, which doesn't stick to dust) is CO. The formation of the CO ice layer provides the raw material from which icy methanol (CH_3OH) and other species may be formed.

Apart from carbon monoxide (CO), the icy species shown in Table 8.1 aren't abundant in interstellar clouds and so are considered to be a result of surface reactions. Just as water ice is formed in surface reactions that attach hydrogen atoms to an oxygen atom on the surface, so methane and ammonia can form in surface reactions that attach hydrogen atoms to carbon and nitrogen on the surface:

$$C \rightarrow CH \rightarrow CH_2 \rightarrow CH_3 \rightarrow CH_4 \text{ (methane)}$$

$$N \rightarrow NH \rightarrow NH_2 \rightarrow NH_3 \text{ (ammonia)}$$

Methanol can form in various ways: the simplest is in surface reactions that attach hydrogen atoms successively to carbon monoxide on the surface:

$$CO \rightarrow HCO \rightarrow H_2CO \rightarrow CH_3O \rightarrow CH_3OH \text{ (methanol)}$$

All of these hydrogen addition reactions can be reversed if the ultraviolet radiation field is strong enough.

Carbon dioxide (CO_2) can be formed in a surface reaction between CO and OH (derived from water):

$$CO + OH \rightarrow H + CO_2 \text{ (carbon dioxide)}$$

Formic acid may be formed in the ice by reactions between OH (from water) and HCO (from formaldehyde) or between O atoms and CH_2OH (from methanol).

8.2 CAN SIMPLE INTERSTELLAR ICES BE A SOURCE OF COMPLEX ORGANIC MOLECULES?

Table 6.3 shows the wide range and great complexity of molecules that have been discovered so far in the interstellar gas. Apart from the fullerenes that are probably formed when carbon grains are eroded in interstellar shocks, these identified molecular species each contain up to about a dozen atoms each. Astronomers call interstellar molecules with more than five atoms per molecule *complex*, and since these detected species nearly all contain carbon as well as other atoms they are classified as *organic*, these larger astronomical molecular species are called *complex organic molecules*, or COMs, for short. Of course, chemists wouldn't regard these molecules as particularly complex, but for astronomers it is reasonable to use the term "complex" because it's not clear how some of them can be formed in the interstellar gas. The chemical variety of the detected COMs is impressive. We give in Box 8.1 a brief description (a *bestiary*, or compendium of "strange beasts") of the types of chemical structures involved. The terminology included in this bestiary will also be useful when we discuss astrobiology in Chapter 10.

Discoveries of interstellar COMs continue to be made. Some of these introduce new and very exciting implications for interstellar chemistry. For example, in 2017, astronomers announced the first detection in interstellar space of a carbon-containing molecule called isopropyl cyanide, $(CH_3)_2CHCN$, in which the carbon backbone is *branched*, rather than straight. Such branching opens up the possibility of even greater complexity in the structure of COMs than we have recognized so far. One year earlier, another complex organic molecule had been found in deep space: propylene oxide, CH_3CHCH_2O, was detected in a cloud of gas and dust called Sagittarius B2, which is located near the centre of the Milky Way galaxy about 25 000 light years away.

Box 8.1   A bestiary of organic molecules: functional groups

Molecules are formed by atoms that bond together. These bonds are either *ionic* or *covalent* bonds. Ionic bonds form when the outermost electron(s) of one atom move permanently to another atom. The atom that loses the electron(s) becomes a positively charged ion, while the other is negatively charged. The atoms are thus joined together by the electrostatic attraction between oppositely charged ions. Alternatively, in a covalent bond the two atoms or ions share one or more electron pairs. The electrons located between the two nuclei are bonding electrons. Covalent bonds occur between identical atoms or between different atoms that have a similar tendency to attract pairs of electrons (called *electronegativity*). Covalent bonds are much more common in organic chemistry than ionic bonds.

The properties of a particular organic molecule are largely dependent on small groups of atoms, two to four, called *functional groups* that exhibit a characteristic reactivity. A functional group generally displays its characteristic chemical behaviour when it is present in a molecule. In fact, organic chemistry is well described by the "functional group approach", where organic molecules are constructed from an inert hydrocarbon skeleton onto which functional groups are attached or super-imposed. Such a simple description works surprisingly well, just because the properties and reaction chemistry of a particular functional group are independent of environment.

Hydrocarbons are organic compounds that are made of only hydrogen and carbon atoms. The simplest configuration is the one in which carbons in the molecules are bonded by single bonds. Molecules (or parts of molecules) containing only carbon–hydrogen and carbon–carbon single bonds are called *alkanes*. Examples of the simplest alkanes are CH_4, C_2H_6, and C_3H_8. Carbon has a valency of four (it can make four bonds) while hydrogen has a valency of one, so one carbon can bind four hydrogens (as in CH_4), or three hydrogens and a bond with another carbon (as in C_2H_6 perhaps better written as CH_3–CH_3), or two hydrogens and two bonds to carbon atoms (as in C_3H_8 perhaps better written as CH_3–CH_2–CH_3). If a hydrocarbon contains a carbon–carbon double bond it is referred to as an *alkene*. The simplest alkene is ethylene, C_2H_4 (or CH_2=CH_2). The detected interstellar molecule propylene, CH_3CHCH_2, is an alkene. A hydrocarbon containing a carbon–carbon triple bond is called an *alkyne*. Acetylene (C_2H_2 or $CH\equiv CH$) is the simplest molecule of this type.

We have already met methanol (CH_3OH). This species contains the simplest possible example of an *alcohol* functional group. In the alcohol functional group, a carbon is single-bonded to an OH group (referred to as a *hydroxyl*). If the central carbon in an alcohol is bonded to only one other carbon, we have a primary alcohol. In secondary alcohols and tertiary alcohols, the central carbon is bonded to two and three carbons, respectively. On this basis, methanol is in a class by itself. Substituting a hydroxyl (OH) with sulfur (*i.e.*, SH) we obtain an analogue of an alcohol called a *thiol*. Just as there are primary, secondary, and tertiary alcohols, there are primary, secondary, and tertiary thiols.

In an *ether* functional group, a central oxygen is bonded to two carbons, as in dimethyl ether (CH_3OCH_3). In *sulfides*, the oxygen atom of an ether has been replaced by a sulfur atom.

Amines are organic derivatives of ammonia, in which one, two, or all three of the hydrogens of ammonia (NH_3) are replaced by organic groups, forming as in alcohols, primary, secondary, and tertiary amines, thus: NH_2R_1, NHR_1R_2, and $NR_1R_2R_3$, where R_1, R_2, and R_3 are organic groups.

A *phosphate ion* is made up of one phosphorus and four oxygen atoms, with the formula PO_4^{3-}. When it is attached to a molecule containing carbon, it is called a *phosphate* group. Many biological organic molecules contain phosphate, diphosphate, and triphosphate groups.

A *carbonyl* group is a chemically organic functional group composed of a carbon atom double-bonded to an oxygen atom, represented by $>C=O$. *Aldehydes* and *ketones* are two closely related carbonyl-based functional groups. In an aldehyde (often called a *formyl* group) the carbonyl group is bonded on one side to a hydrogen, $H-C=O$, and on the other side to a carbon. In ketones the remaining two bonds of the carbonyl group, $>C=O$, are to other carbon atoms or hydrocarbon radicals. The nitrogen analogue of a carbonyl group, where C is replaced by N, is called the *imine* group. The *carboxylic acid derivatives* are a family of closely related functional groups, in which the carbonyl carbon is bonded on one side to a carbon (or hydrogen) and on the other side to a *heteroatom*. In organic chemistry, heteroatoms are oxygen, nitrogen, sulfur, or one of the halogens. This is what distinguishes carboxylic acid derivatives from aldehydes and ketones. The eponymous member of the family is the *carboxylic acid* group, in which the carbon belonging to the carbonyl is bonded to an OH group. The carboxyl group is frequently indicated as $-COOH$. When the hydrogen is lost from the hydroxyl group we obtain a *carboxylate*. As the name implies, carboxylic acids are acidic, meaning that they are readily deprotonated forming a *base*.

Amides are derived from the carboxylic acid in which an OH group is substituted by an amine group, $-NH_2$. So, amides contain the $-CONH_2$ group. We may think of amides as carbonyls bonded to an amine. If the carbonyl is bonded to an alcohol (with the central carbon belonging to both groups) we obtain an *ester*. When the oxygen in the alcohol part is swapped with a sulfur atom, the new compound is called a *thioester*. A phosphate group when linked to a carbon atom is called a *phosphate ester*.

In a *nitrile* group, a carbon is triple-bonded to a nitrogen, represented by $-C\equiv N$. Nitriles are also often referred to as *cyano* groups.

Hydrocarbon and *carbohydrate* sound similar but they are indeed very different types of molecule. Hydrocarbons are composed only of carbon and hydrogen, while carbohydrates also contain oxygen. Living organisms metabolize carbohydrates for energy, so it is not surprising that sugars are carbohydrates. The chemistry of carbohydrates is complicated by the fact that there is an alcohol on almost every carbon. They may exist in either a straight chain or a ring structure. Ring structures incorporate two additional functional groups: the *hemiacetal* and *acetal*. A hemiacetal is a carbon connected to two oxygen atoms, where one oxygen is an alcohol and the other is an ether. Acetals are compounds characterized by the grouping carbon atoms being bonded to two ethers. They may be synthesized by heating aldehydes or ketones with alcohols. Carbohydrates are broadly classified into *monosaccharides*, *disaccharides* and *polysaccharides*. Monosaccharides and disaccharides are otherwise known as *sugars*.

(*continued*)

Box 8.1 *(continued)*

There are no standard rules for identifying *sugars*. Simple sugars (also called monosaccharides) have the chemical formula $C_n(H_2O)_n$, where n is at least 3. *Glycolaldehyde* ($C_2H_4O_2$) is the smallest possible molecule that contains both an aldehyde group and a hydroxyl group. Although it conforms to the general formula for carbohydrates, it is not generally considered to be a saccharide. *Glucose* ($C_6H_{12}O_6$) is a monosaccharide, widespread in Nature (*e.g.*, it is the sugar found in fruits), either as a single species, or abundantly present in complex compounds. A disaccharide is a combination of two monosaccharide molecules by the removal of a molecule of H_2O. The polysaccharide is made up of hundreds or even thousands of glucose molecules. *Ribose* has the formula $C_5H_{10}O_5$, and is an important piece of RNA. A derivative of ribose, *deoxyribose* also has five carbon atoms but an oxygen is lost ($C_5H_{10}O_4$). It is a key component of DNA, that together with RNA carry the genetic code of living organisms. Three carbon monosaccharides are called *trioses*, four carbon are called *tetroses*, five carbon are called *pentoses*, six carbon are *hexoses*, and so on. The number of carbon atoms in a molecule is used in the suffix of carbohydrate naming.

Finally, *amino acids* are organic compounds composed of nitrogen, carbon, hydrogen and oxygen. They are a type of organic acid that contains a carboxyl functional group (–COOH) and an amine functional group (–NH₂), as well as a side-chain group that is specific to the individual amino acid. Amino acids, often referred to as the *building blocks of proteins* (in that case they are called *proteinogenic amino acids*), are compounds that play many critical roles in our body, which needs 20 different amino acids to grow and function properly. *Peptides* are composed of two or more amino acids joined through amide formation involving the carboxyl group of one amino acid and the amino group of the next. The chemical bond between the carbon and nitrogen of each amide is called a *peptide bond*.

Propylene oxide is particularly interesting because it is a *chiral* molecule, meaning that it exists in non-super-imposable forms that are mirror images of one another, in the same way that our left and right hands cannot be precisely super-imposed on each other. The two forms of a chiral molecule, called *enantiomers*, have identical physical and chemical properties, but the way they interact with other chiral molecules may be different (for example, the interaction of a left foot in a right shoe is different from that of a left foot in a left shoe). The two forms of propylene oxide are shown in Figure 8.2.

Every living thing on Earth uses one *and only one* "handedness" of many types of chiral molecules. This trait, called *homochirality*, is an essential feature of terrestrial biochemistry, and has important implications for many biological structures, including DNA's

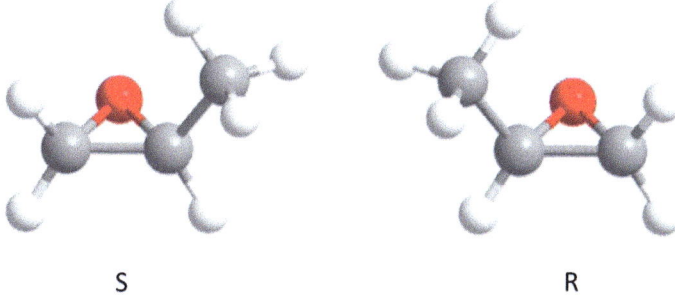

S   R

Figure 8.2   The structures of two molecules that are mirror images of each
other; these are *enantiomers*. The two molecules cannot be super-
imposed on each other, and are conventionally labelled *R* (for rec-
tus, or right) and *S* (for sinister, or left). The molecule illustrated
is propylene oxide.

double helix. As such, the discovery that chirality exists far outside
our solar system is one with huge consequences. We'll meet such
organic species important in *pre-biotic chemistry and biochemistry*
in Chapter 10, where we'll discuss the way larger species such as
amino acids, the basic constituents of proteins, or sugars naturally form in space before being incorporated into asteroids and
comets, and later deposited on young planets. Such topics are
now considered to be part of the emerging field of *astrobiology*.

COMs are found in a variety of locations in interstellar space.
These are briefly described in Box 8.2. The physical conditions
in these regions cover a wide range, so the formation of COMs
must be a robust mechanism, and not very sensitive to precise
conditions.

As we discussed in Chapter 6, many of the smaller molecular
species can be made in reactions in the gas in interstellar clouds
in networks of reactions of the kind that we described in that
chapter. When these ideas are tested by calculations, it is clear
that these networks are capable of providing abundances of
smaller species that are comparable to those observed in various
locations of interstellar space, so those ideas are regarded as successful; *i.e.*, many of the smaller detected molecular species can
be made in the interstellar gas. However, as we noted earlier, the
larger molecular species present some problems for chemistry
in interstellar gas. It's not clear how many of these species can

Box 8.2   Some regions of the Milky Way in which COMs have been detected

We'll discuss star formation in more detail in Chapter 9. But, summarizing briefly, it involves the gravitational collapse and fragmentation of interstellar gas clouds. The gravitational collapse is opposed by the pressure of the gas, by the resistance provided by the embedded magnetic field, by turbulence in the gas, and by the rotation of the cloud. We'll see in Chapter 9 that chemistry during these dynamical processes produces molecules that are effective radiators of energy, even at low temperatures, so that gravitational energy released as heat during the collapse is radiated away and the clouds remain cool during the collapse. Without these coolants, the gravitational energy released in the collapse would heat the cloud, increase the gas pressure, and stop the collapse. (Remember how a bicycle pump gets hot when you pump up a tyre? You are doing work on the air inside the pump, and some of the energy you exert appears at heat. In star formation, gravity is doing the work of compressing the gas, and some of that gravitational energy appears as heat.) The end-point of this collapse process is the formation of a star.

Therefore, we expect to find the densest gas associated with very young stars, still embedded in the gas from which they were formed. Initially, these stars are heated by the release of gravitational energy that occurs at an increasing rate as the star is forming. In this phase, the star is a powerful source of infrared radiation, and heats the gas in its immediate vicinity to temperatures of a few hundred degrees above absolute zero. Eventually, the compression of the gas by gravitational forces creates conditions in the interior of the proto-star that enable thermonuclear processes to switch on, and the star rapidly heats up and becomes a powerful radiator in the visible and ultraviolet. At this stage the newly-formed stars generate jets, outflows and winds that impinge on the progenitor clouds, causing shocks and turbulence. The densest gas found in association with the youngest stars is thus subjected to a variety of energetic processes that unavoidably cause the chemistry to respond.

This picture implies that a newly-formed star should be initially embedded in a region of warm, very dense gas, glowing in the infrared. This is indeed what is observed. These are tiny regions, around one percent or so of the size of a typical molecular cloud. The striking result of decades of observations is that the molecular species detected in the gas close to very young stars are exceptionally complex and are present in substantial abundances. These tiny molecular regions are called *hot cores* when associated with massive and powerful proto-stars, and *warm cores* (or sometimes *hot corinos*) when associated with less massive and less powerful young stars. The temperatures in these regions are typically a few hundred Kelvin, and the number densities of hydrogen molecules are typically ten or a hundred million per cubic centimetre, so they are enormously dense compared to the interstellar dark clouds in which the collapse process began, and they are rather warmer, too. The enormous variety of molecular species and their remarkable complexity found in the gas of hot cores is astounding. A hot core known as Sagittarius B2 in

Figure 8.3 Image of the centre of the Milky Way galaxy. This false-colour infrared image shows the infrared radiation from dust in the galactic centre clouds that are heated by powerful radiation from young massive stars (credit: NASA/JPL-Caltech).

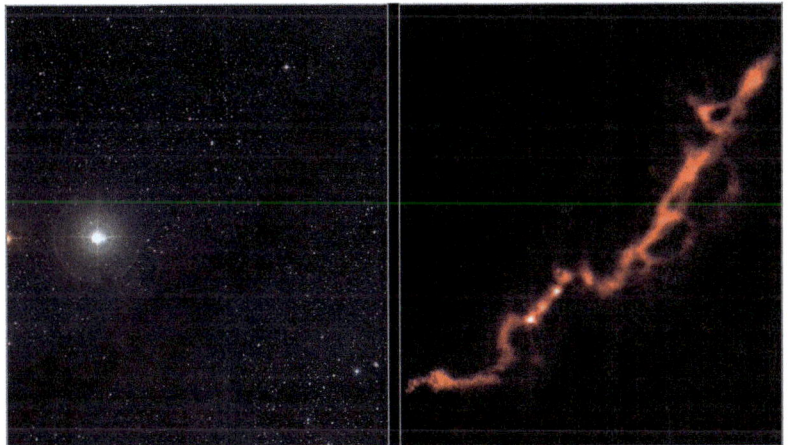

Figure 8.4 Image of TMC-1. The Taurus Molecular Cloud is about 450 light years from Earth, and because this is relatively close, TMC-1 is a favourite object for astronomers to observe. The image compares how the cloud appears dark when observed in visible light (left) with the appearance in the infrared emitted by warm dust (right) (credit: ESO/APEX (MPIfR/ESO/ OSO)/A. Hacar *et al.*/Digitized Sky Survey 2. Acknowledgment: Davide De Martin).

(*continued*)

Box 8.2   (*continued*)

the centre of the Milky Way galaxy holds the record for the discovery of most COMs in one object (see Figure 8.3).

Hot and warm cores associated with newly-forming stars are not the only sources of large molecules. They are also found in other types of interstellar region. The extended outflowing envelopes around cool stars and planetary nebulae, discussed in Chapter 4 as sources of dust grains, are also places where COMs are found. A cold interstellar cloud, the Taurus Molecular Cloud-1 (TMC-1, see Figure 8.4), has been shown to be exceptionally rich in large carbon-chain species.

An extremely complex chemistry is also found in dense cores of gas inside gravitationally collapsing clouds even before a proto-star appears. These so-called *pre-stellar* cores are cold, since no star has yet appeared, and are found to be rich in COMs. Circumstellar discs appear during the star-formation process and are the locations of comets, meteorites, and planets. Complex molecules have been detected in the discs, comets and meteorites. COMs have also been found in the so-called *Orion Bar* (not, unfortunately, a source of interstellar refreshments). The Orion Bar (see Figure 8.5) is a place where intense radiation from a very massive star in the Orion constellation impinges on a nearby dense gas cloud, and from Earth we happen to look at this region edge-on, so it appears to be elongated.

Figure 8.5   Image of the Orion Bar. Powerful winds from massive stars interact with a nearby molecular cloud. The view from Earth shows this interface edge-on, so that it appears as a bar. It is a region where intense radiation fields are mixing with a dense and dusty interstellar cloud. Such regions show a very active chemistry [credit: NASA/C.R. O'Dell and S. K. Wong (Rice University)].

be made in the gas, and when some ideas for COMs chemistry in the interstellar gas are proposed and tested by detailed calculations, we find that these chemical networks in the gas cannot make enough of the product molecules.

What's to be done? What other ways of forming COMs are available? Table 8.1 shows us that interstellar ices contain, in rather simple molecules, significant amounts of the important elements carbon, nitrogen, and oxygen (*etc.*) that are found in the COMs. Is it possible that a chemistry using interstellar ice as the source materials can successfully make large abundances of COMs in interstellar space? It seems clear that we can't afford to ignore this vast reservoir of carbon, nitrogen, and oxygen contained in simple molecules in interstellar ices.

A huge amount of laboratory work and theoretical modelling now supports the idea that it is possible to convert the simple molecules contained in interstellar ices into quite complex species in the interstellar gas. We'll describe some of this work in the next section.

8.3 'COOKING' SIMPLE ICES MAKES COMPLEX MOLECULES IN INTERSTELLAR SPACE

All of the interstellar locations where large molecules are detected are of high density. This means that the freeze-out time is short and almost all species in the interstellar gas (apart from hydrogen molecules and helium atoms) will stick to the surface of dust grains in the form of the simple interstellar ices described in Section 8.1. Therefore, most of the atoms required to make the COMs are in the simple ices. This suggests that interstellar ices may be involved in the chemistry of COMs: could the way to make complex molecules in space involve 'cooking' the simple ingredients – water, carbon monoxide and dioxide, methane and ammonia – that are present in the ices? This cooking now seems to be a very likely scenario.

But what is meant by 'cooking' in the interstellar medium? That is, what kind of chemical processing can occur to make the COMs? The types of region where COMs are detected, described in Box 8.2, are all rather different. For example, some are very dark while others may be exposed to radiation, so the processing may not be the same for all situations. However, the ice contains much of the material that could be turned into COMs, so this seems the best place to start.

An imaginative proposal was put forward about a decade ago by Robin Garrod, Susanna Widicus Weaver and Eric Herbst to account for the astounding variety of large molecules found in *hot cores*, very dense and small clouds of warm gas near to very young massive stars (see Box 8.2). They noted that the ices in the cores were subject to a weak ultraviolet radiation field that broke down a few of the molecules into simpler and much more reactive forms. For example:

$$H_2O \rightarrow OH + H \rightarrow O + H + H$$

$$CH_4 \rightarrow CH_3 + H \rightarrow CH_2 + H + H \rightarrow CH + H + H + H$$

$$NH_3 \rightarrow NH_2 + H \rightarrow NH + H + H$$

$$CH_3OH \rightarrow CH_3O + H$$

etc. The new species: OH, CH, CH_2, CH_3, NH, NH_2, and CH_3O are molecules lacking one or more hydrogen atoms. They are *radicals*, and are highly reactive.

Before the proto-star formed, the ices were very cold, and these radicals were trapped in the ice and could not move. But when the newly-forming star formed and irradiated the core, then the temperature of the ice warmed up from about ten degrees to about thirty degrees above absolute zero. At these higher temperatures, these new species become mobile in the ice and can react together to form new species in simple addition reactions, for example:

$$CH_3 + OH \rightarrow CH_3OH \text{ (methanol)}$$

$$CH_3 + NH_2 \rightarrow CH_3NH_2 \text{ (methylamine)}$$

$$CH_3 + CH_3O \rightarrow CH_3OCH_3 \text{ (dimethyl ether)}$$

(all of which are observed COMs). Sometimes two stages of reaction might occur, for example:

$$CH_2 + OH \rightarrow CH_2OH \text{ (hydroxymethyl radical)}$$

$$CH_2OH + CH_3 \rightarrow CH_3CH_2OH \text{ (ethanol)}$$

Or, in more examples:

$$CH_3 + CH_2 \rightarrow CH_3CH_2 \text{ (ethyl radical)}$$

$$CH_3CH_2 + CH_3 \rightarrow CH_3CH_2CH_3 \text{ (propane)}$$

and:

$$CH_3 + CH \rightarrow CH_3CH$$

$$CH_3CH + CH_2 \rightarrow CH_3CHCH_2 \text{ (propylene)}$$

These products are also observed COMs.

A very large network of chemical reactions of this type can be generated from a fairly small number of radicals; for example, with, say, ten different radicals over one hundred different product molecules may arise. So the network will, in principal, create a huge list of possible products, some of which are familiar from school chemistry. Very many of these predicted molecular species are in fact COMs detected in the interstellar gas. These products will also be trapped in the ice. Assuming that this solid-state chemistry occurs in space, then the radiation from the evolving proto-star will eventually cause the ice temperature to rise to the point (more than one hundred degrees above absolute zero) at which the ice is rapidly evaporated, so that at least some of the products enter the gas phase to be detected by astronomical observation. Other molecules, particularly the largest and most complex molecules, may stick to the dust grains and be incorporated into meteorites, comets, and planets.

Of course, there are many assumptions in a model of this type of COM chemistry. The details of the solid-state chemistry occurring in the ice are largely unknown. But the predictions of the model are very impressive: a large network of possible reactions arises from a rather limited number of radicals, and a huge variety of COMs can be formed in this network and released to the interstellar gas when the ices evaporate.

This method of making COMs is plausible, even though many details are lacking. However, given the variety of types of interstellar location in which COMs are found, it's unlikely that this is the only way in which simple ices can be cooked to make more complex species. The simple molecules frozen on dust surfaces are preserved for millions of years, giving many opportunities for cooking. This can be driven by fast charged particles, by energetic radiation, in particular by ultraviolet photons and by X-rays, as well as simply by warming up the ice. This cooking (astrochemists actually use the term *processing*) may be the key to revealing how simple molecules in space evolve into greater and greater complexity, eventually including the building blocks of biology. Processing provides firstly the activation of molecules initially present in the ice, allowing the development of an important chemical reactivity when the dust temperature reaches about thirty degrees above absolute zero, permitting radicals to diffuse through the ice and giving the start to a rich solid-state chemistry. Reactive species such as ions and radicals diffuse within the bulk of the ice, and ultimately re-combine to form new and more complex molecules. As we'll see in the next section, there is a great deal of support for these ideas about processing of ices from chemical experiments carried out in laboratories on Earth.

Observations show the presence of some COMs even in prestellar cores, that is, before a young star has 'switched on' so the dust temperature is less than thirty degrees. This may suggest that solid-state chemistry can take place even at these lower temperatures and that chemical energy released in those reactions is enough to warm the grains and kick the product molecules off the ice into the gas. Alternatively, it may mean that chemical conversion to greater complexity occurs by reactions in the gas, but we've seen already that there are problems with this approach. So the processing of ices still seems the best bet for COMs production, and as we'll see in the next section is supported by the results of experiments in the laboratory. The possible evolution of ices during star formation and formation of complex molecules is shown in Figure 8.6.

This area of astrochemistry is still very much a "work in progress", and in looking for answers, scientists are frequently generating more questions. The growth of molecular complexity is difficult to trace in astronomical regions, because the dominant simple ice species hides the minor signatures from COMs. How then can we hope to capture a glimpse of the ongoing chemistry?

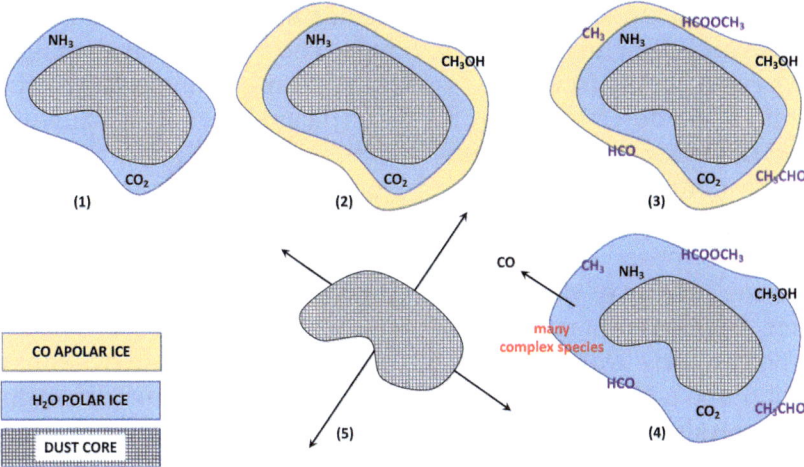

Figure 8.6 Evolution of ice mantles in star-forming regions. Observations of dust grains with icy mantles have indicated the following sequence of evolutionary stages. In stage (1), a water ice mantle is deposited on the dust grain, and ammonia and carbon dioxide are formed. In stage (2), a carbon monoxide layer is deposited. Some of this CO is converted to methanol (CH_3OH). More complex chemistry occurs in stage (3), in which molecules as complex as methyl formate ($HCOOCH_3$) and acetaldehyde (CH_3HCO) appear. The effects of heating and irradiation create radicals that promote a wider chemistry in stage (4). Shocks or other processes in stage (5) may remove mantles completely or partially and eject products into the gas.

The answer to such fundamental astrochemical questions can only be obtained by means of laboratory experiments. During an experiment, researchers try to simulate conditions in space to understand what chemical reactions could occur there. Temperature, radiation, vacuum and composition of laboratory simulations are thus crucial. As you might guess, to replicate all these complications in a laboratory is not an easy task.

8.4   LABORATORY ASTROCHEMISTRY: OUTER SPACE IN A BOX

Ideas such as those we have discussed in the previous section about the processing of simple interstellar ices to make molecules of greater complexity are obviously important. However, it is also necessary to test these ideas by laboratory experiments carried out under conditions that are as close as possible to those prevailing in interstellar clouds. These are highly demanding experiments, requiring great skill in many laboratory techniques. We describe in Box 8.3 the typical kinds of experiments and how they are carried out.

The main conclusion arising from these experiments is that they support very strongly the idea that processing of simple ices can make other – more complex – molecules, and that the processing may be of various kinds, involving irradiation by ultraviolet, X-rays, fast particles, *etc.*

Some early work carried out over a decade ago by Naoki Watanabe and colleagues in Japan with Valerio Pirronello from Sicily, Italy, established the origin of the largely unprocessed ice. Since the interstellar gas is largely molecular hydrogen and carbon monoxide, why does the simplest interstellar ice have the composition shown in Table 8.1? In their experiments, Watanabe and colleagues made ice consisting only of water and carbon monoxide, and irradiated it with hydrogen atoms and with ultraviolet light to represent the situation in the interstellar gas. The chemistry of the ice evolved in the experiment, and new molecules appeared. These were carbon dioxide (CO_2), formaldehyde (H_2CO), methanol (CH_3OH), and formic acid ($HCOOH$), precisely the mixture of molecules that is found in simple ices. Evidently, the ideas of processing by H atoms and ultraviolet radiation seem to be well established. Even more impressive is the fact that the relative abundances of these molecules to water obtained from the experiments is in quite reasonable agreement with the relative abundances observed in the interstellar ices. From their experiments, Watanabe and colleagues were able to establish precisely how this chemical evolution occurred. So we can conclude that the nature of simple interstellar ices detected in interstellar clouds is well understood, and the main influences on this ice are hydrogen atoms and ultraviolet radiation.

The next step, therefore, was to understand the chemical evolution of molecules from the simple ices into more complex molecules. The main drivers of this chemical evolution that have been explored experimentally are ultraviolet radiation, X-rays, fast electron and fast atom irradiation. Karin Öberg (then at Leiden, Netherlands) and colleagues showed

Box 8.3   The astrochemical laboratory

The experiments carried out in a solid-state astrochemistry laboratory are designed to mimic chemical reactions that take place naturally in the interstellar gas or on the surface of a dust grain in the harsh environment of space. We describe here the experimental set-up used by many investigators to study evolving interstellar ices. As we have explained in Sections 8.2 to 8.4, such ices contain familiar molecules, such as water, carbon monoxide, carbon dioxide, methanol, ammonia and methane, that – when exposed to various forms of high-energy processing – form new and more complex chemicals that did not previously exist in the ice. These experiments have formed many intriguing organic compounds, including amino acids and sugars.

Laboratories use vacuum equipment to pump out most of the air from a *vacuum chamber*, creating a volume of very high vacuum. Figure 8.7 shows a picture of an astrochemical laboratory at the Observatory of Palermo.

Ice of the desired composition is deposited on a very cold surface, representing the dust grain surface, by aiming a beam of the appropriate molecules at the cold surface. The apparatus is equipped with windows through which the ices are exposed to energy sources. The ices are analyzed by

Figure 8.7   An astrochemistry laboratory apparatus. The LIFE laboratory (*Light Irradiation Facility for Exochemistry*) is at the Observatory of Palermo. The ultrahigh vacuum chamber is at the centre. The ultraviolet lamp (also shown in Figure 8.8) is illuminating a sample through a window that is transparent to ultraviolet light. The cryostat with the downwards pointing *cold finger* is located at the top of the chamber. The grey box on the right of the picture is the infrared spectrometer, while the smaller grey box on the left is the mass spectrometer. The pumping system and alternative energetic sources other than ultraviolet radiation (such as X-rays and energetic electrons) are not shown in this picture.

(*continued*)

Box 8.3   (*continued*)

infrared, ultraviolet, and mass spectrometers monitoring the ices before, throughout, and after irradiation, and during any subsequent warming and evaporation of the ice. When these ices are warmed after irradiation, volatile materials evaporate into the gas phase, leaving behind an organic residue on the sample head. The analysis of these new materials is made outside the chamber using separate devices.

To effectively simulate the conditions in space, the chamber should operate in ultrahigh vacuum, a vacuum regime characterized by pressures that should be approximately one thousand billion times lower than the pressure at ground level on Earth. There is no single vacuum pump that can operate all the way from atmospheric pressure to ultrahigh vacuum. Instead, a series of different pumps is used, according to the appropriate pressure range for each pump. A turbomolecular pump is designed to reach ultrahigh vacuum. It consists of a rotor and stator blades, providing transfer of impulses from the rapidly rotating blades to the gas molecules being pumped. Molecules are adsorbed by the blade material and leave the blades after a certain period of time. To work effectively, the speed component added by the blades must not be lost through mutual molecular collisions. In other words, the distance that a molecule travels before colliding with another one must be greater than the blade spacing. The gas captured is compressed to the pressure level of pre-vacuum provided by a backing pump. Typically, this backing pressure is provided by another kind of pumping system, such as a scroll vacuum pump. These devices use two interleaved spiral-shaped scrolls, one being fixed while the other orbits eccentrically without rotating, thereby trapping and compressing pockets of fluid between the scrolls. Other kinds of pump also exist, such as cryopumps, devices that trap gases and vapours by condensing them on a cold surface, but they are only effective for some gases.

Once the chamber reaches the desired pressure, the gases to be condensed are injected in a mixing chamber, and then allowed to flow towards the sample holder on which the ice is formed, located at the tip end of a cryostat, a device by which temperature can be maintained at a low level. In general, it is called a *cold finger* (because of its resemblance to a finger!); the finger has a chamber where a coolant fluid can enter and leave. To reach the extremely low temperatures (less than twenty degrees above absolute zero) of deep space, the cold finger is cooled by cold helium gas that has been cooled by its expansion inside a cylinder containing a moving piston (as in familiar refrigeration systems).

There are several energy sources through which the ices are processed. The most widely used is ultraviolet light, commonly simulated by inexpensive and easy-to-use microwave-powered hydrogen discharge lamps. These consist of a slim tube containing hydrogen gas at low pressure with an electrode at each end. Setting a high voltage (a few thousand Volts) makes the tube light up with a bright pink glow, which is a small part of the hydrogen emission spectrum (Figure 8.8). Most of this emission is invisible because it is in the ultraviolet, the part of the spectrum that can cause chemical transformations in the ice.

In some experiments the intention may be to irradiate the ice with X-rays; these are generally produced through electronic impact. In this source, electrons emitted from a filament made of specific materials are focused onto a

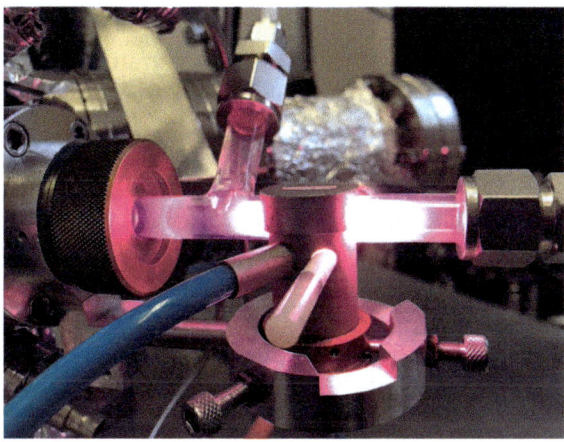

Figure 8.8   The ultraviolet lamp at work. The pink glow is the part of the spectrum of the lamp's emission to which human eyes respond. Most of the emission is in the ultraviolet range. The ice sample in the chamber is exposed to the lamp's radiation through a window that is transparent to ultraviolet radiation.

metal target. The absorbing target material re-emits energy as characteristic X-ray emission lines lying on top of a continuum of X-radiation. Such kinds of devices are compact and portable, but their X-ray output is low. For more intense beams and better optical performance, synchrotron storage rings are needed. These radiation sources are large machines (about the size of a football field) that accelerate electrons to almost the speed of light. The electrons are deflected by magnetic fields and create extremely bright radiation at all wavelengths from the infrared, through the visible and ultraviolet, and into the X-ray region. This radiation is channelled down beam lines to experimental workstations where it is used for research (Figure 8.9).

Particles such as electrons, protons, and heavy ions are also efficient ice-processing agents. Electrons are produced through an instrument called an *electron gun*, that in its most basic form consists of a heated cathode (such as the one in television sets) creating a stream of electrons through thermionic emission. Electrodes generate an electric field, which focuses the beam towards an anode. A large voltage between the cathode and anode accelerates the electrons.

Finally, the ice modifications are analyzed through infrared and mass spectroscopy. The infrared spectroscopy identifies molecules based on their functional groups (see Box 8.1). When infrared radiation hits a molecule, the bonds in the molecule absorb the energy of the infrared photon, and respond by vibrating. The most common types of vibrations are bending, stretching, rocking, and scissoring. This gives rise to an infrared spectrum in which each 'wiggle' probes a specific functional group or molecule. The mass spectrometer is an instrument used to measure the mass of ionized atoms or other electrically charged particles. Mass spectrometers are used in physics, geology, chemistry, biology and medicine

(*continued*)

Box 8.3 *(continued)*

to determine compositions, to measure isotopic ratios, and for detecting leaks in vacuum systems. The principles on which this instrument works are very simple: if something is moving and you subject it to a sideways force, instead of moving in a straight line, it will move in a curve. In the case of a moving ball, if we know its velocity and the magnitude of the deflecting force, by measuring the deflection we can infer the ball's mass: the less the deflection, the heavier the ball. We can apply exactly the same principle to atoms and molecules. Electrically charged particles are affected by a magnetic field that provides a force perpendicular to their motions. Thus, the first step in the measurement is knocking one or more electrons off an atom to give a positive ion. The ion is then deflected by a magnetic field according to its mass. The more the ion is charged, the more it gets deflected. Then, more correctly what a mass spectrometer measures is the mass-to-charge ratio.

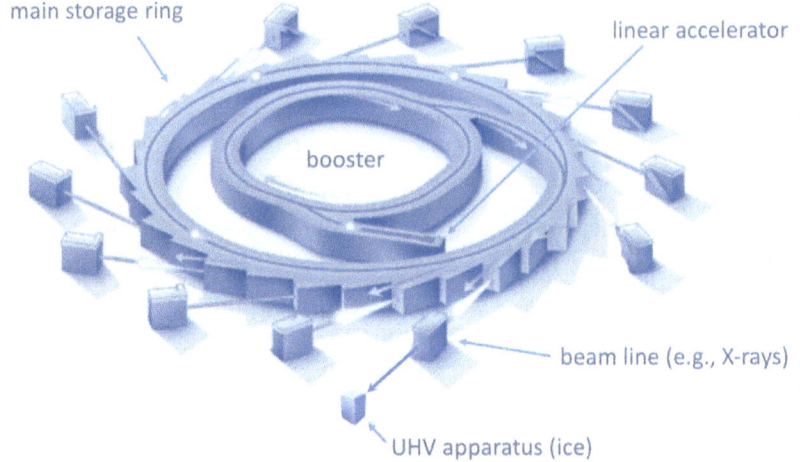

main storage ring

linear accelerator

booster

beam line (e.g., X-rays)

UHV apparatus (ice)

Figure 8.9   A schematic view of a synchrotron apparatus. Electrons are initially accelerated along a straight line by powerful radio frequency fields and then injected into the booster ring where they are further accelerated to energies up to a thousand times higher. When the electrons have enough energy to produce light, they are transferred to the storage ring. Windows are placed along the storage ring to allow the light to escape along the beam lines, at the end of which the control cabins are located. In the cabin's researchers can 'shape' and measure the radiation. Finally, experimental facilities (such as the one shown in Figure 8.7) channel radiation onto the sample.

that ices of methanol, or methanol together with water, when irradiated with ultraviolet radiation generated a large variety of organic molecules, including ethane (C_2H_6), acetaldehyde (CH_3CHO), ethanol (C_2H_5OH), dimethyl ether (CH_3OCH_3), methyl formate ($HCOOCH_3$), glycolaldehyde ($HOCH_2CHO$), and ethylene glycol ($(CH_2OH)_2$). All of these molecular species have been detected in interstellar gas. In the experiments, the production of these species was found to be efficient enough to account for the observed interstellar abundances of these COMs. Remarkably, Y. J Chen, Cesare Cecchi-Pestellini and colleagues found that if soft X-rays are used to irradiate methanol ice instead of ultraviolet, the range and variety of product molecules is quite similar to that obtained from either ultraviolet or fast electron irradiation. It seems that the actual method of processing doesn't matter too much; we still get a similar range of products regardless of how the ice is cooked. The important result appears to be that we have to energize the ice in some way, and once that is done, a similar range of products emerges.

8.5 CONCLUSIONS

The observational information about complex molecules in space can seem quite complicated. The ices are fairly simple, but the impressive variety of complex molecules discovered to exist in many different types of interstellar region is almost bewildering. However, as we have shown in this chapter, we can put together a straightforward story about interstellar ices and complex molecules in interstellar gas. While some loose ends remain to be tied up, the conclusions of this story are strongly supported by many laboratory experiments that aim to replicate interstellar chemistry. Dust grains play a key role throughout this story. Let's summarize the narrative here.

Ices form on the surfaces of dust grains in dark clouds from which starlight is to some extent excluded because of the extinction caused by those same dust grains. Some of the molecules in the ice are formed in surface reactions and retained at the surface. Water ice is the most important of these species, and is formed when oxygen atoms collide with the surface of a dust grain and add hydrogen atoms to form water molecules. Carbon and nitrogen atoms colliding with dust grains are similarly hydrogenated to form methane and nitrogen. Carbon monoxide, the most abundant molecule in interstellar gas after hydrogen, also sticks to dust grains; some of these carbon monoxide molecules are hydrogenated to form formaldehyde and methanol, while others on the surface react with hydroxyl radicals to form carbon dioxide.

These simple ices are the feedstock from which the observed complex molecules are formed. Complex molecules are found in much denser gas than dark clouds, so most of the interstellar gas in these regions – apart from hydrogen – has condensed into ices. These ices are processed by one or more sources of energy, such as ultraviolet radiation, X-rays, fast electrons or other particles, which are available in interstellar space. Experiments confirm that the processing of simple ices produces a wide range and variety of molecules, and that the precise energy source seems to be of minor importance. All we need to do to stimulate the formation of COMs from the simple ices is energize the system.

Evidently, dust grains are absolutely essential players at all stages in this story. They provide the surfaces on which reactions occur and product molecules are retained, so that simple ices can form. These simple ices are the material from which complex molecules are created in energetic processing. Chemical complexity in the interstellar gas depends on the existence of dust grains.

Making Stars and Planets From Interstellar Gas and Dust

9.1 INTRODUCTION

The formation of stars and their planetary systems is among the topics of greatest interest in modern astronomy. Our Earth and the Solar System were born approximately 4.5 billion years ago, and our knowledge of that ancient event is sparse. Fortunately, star formation in the Milky Way galaxy is an ongoing phenomenon, and with modern astronomical instrumentation we can observe the process in many locations and in various states of evolution. Important clues about the genesis of our planetary system lie hidden behind the veil of the dusty star-forming molecular clouds. Dust clouds scatter visible light, but let infrared light pass through relatively unimpeded. Near-infrared light, at wavelengths of a few microns, can pierce through the dusty veil and provide a view of the birth of other stars and stellar systems, acting as a time machine to capture an echo of the events that created the Sun and our own Solar System.

The star-formation process is important not only for the star itself and its immediate environment, but also affects its parent

Dust in Galaxies
By David A. Williams and Cesare Cecchi-Pestellini
© David A. Williams and Cesare Cecchi-Pestellini 2020
Published by the Royal Society of Chemistry, www.rsc.org

galaxy as a whole. Star formation converts the galactic reservoir of interstellar gas into stars, flooding interstellar space with infrared, optical, ultraviolet and X-radiation, and – through supernovae explosions – maintaining a low-density million-degree background gas as the stage on which a galaxy evolves. Moreover, through stellar winds and explosions this process controls the buildup of heavy elements and dust in the Universe, which are responsible for the creation of planetary environments in which life might be possible.

A molecular cloud can be in an approximate equilibrium between the forces acting to expand the cloud and those trying to contract it, primarily gravity. However, if this equilibrium is perturbed, either by an external force, such as a supernova explosion, or by the cloud becoming sufficiently massive, these huge clouds collapse under their own weight to form stars, and in doing so are subjected to immense changes in their physical properties. The typical number density of an interstellar cloud gas is a few hundred hydrogen atoms per cubic centimetre, while its temperature is approximately ten degrees. Turning these conditions into those of a star like the Sun requires an increase in number density by more than twenty orders of magnitude, and in central temperature by a factor of about a million. The birth of a star is a violent and chaotic event, with gas flowing in and being ejected outwards at speeds up to hundreds of kilometres per second. Indeed, when the cloud that generated the Solar System contracted under its own gravity and our *proto-Sun* formed in the hot dense centre, the process was locally opposed by thermal, turbulent, and magnetic pressures, by dynamical outflows, and – since the parent cloud was rotating – by angular momentum effects. Because of the competition of all these processes, a part of the contracting cloud formed a swirling disc called the *solar nebula*. Circumstellar discs are an inevitable consequence of angular momentum conservation during the formation of a star through gravitational collapse. Initially, discs rapidly funnel material onto the star but, as the surrounding molecular core is used up or otherwise disperses, the accretion rate decreases, and only a small amount of the original material persists in the disc. You can read more about angular momentum in Box 9.1.

Box 9.1 Angular momentum

Angular momentum is the rotational counterpart of linear momentum, and both of them deal with how quickly something is moving and how difficult it is to change that motion. The linear momentum of an object is basically "the mass of the object in motion", and describes the quantity of motion of that object. Formally, we define the linear momentum as a vector quantity (*i.e.*, having a direction in space) given by the product of an object's mass and its velocity vector, tying velocity and mass into one quantity. A body's momentum is thus always in the same direction as its velocity vector. It might not be obvious why this is useful, but momentum is important because its total amount in a collection of objects never changes, if no external forces are acting. This principle of *momentum conservation* is one of the most powerful laws in physics. This is particularly evident in the collision between two particles: the total momentum of the two particles before the collision is equal to the total momentum of the two particles after the collision. That is, the momentum lost by one object is equal to the momentum gained by the other one.

Angular momentum is the rotational equivalent of linear momentum; it measures the quantity of rotational motion. It is the product of the moment of inertia of a body (or *inertia*) and the angular velocity. Inertia is a body's resistance to change in speed and increases with the mass of the body and the distance of that mass from the axis of rotation. Angular velocity is the rate of velocity at which an object or a particle is rotating around a specific point. The direction of the angular velocity vector is perpendicular to the plane in which the rotation takes place. The angular momentum has the same direction as the angular velocity, and it is thus perpendicular to the plane of rotation. For instance, the angular momentum of a merry-go-round is directed towards the sky, perpendicularly to the ground, while for a top it precesses slowly around a vertical axis through the top point of support while the top spins rapidly about its own axis. Just like its linear counterpart, the angular momentum satisfies a conservation principle.

To change the linear momentum of a body we need to apply a force acting over a finite time interval. What causes a change in angular momentum? To answer this question, we need to introduce the *torque*, defined as the tendency of a force to rotate a body to which it is applied. This quantity is specified with regard to the axis of rotation, and it is equal to the magnitude of the component of the force vector lying in the plane perpendicular to the axis, multiplied by the shortest distance between the axis and the direction of the force component. Thus, the larger the distance the weaker the force we need to apply to a body to make it rotate. This is reason why doorknobs are placed as far as possible from the hinges!

A star and its planetary system are born from the collapse of dense interstellar cloud. Such clouds are generally very huge. Let's consider a portion of a slowly rotating collapsing cloud initially about one light year wide. In the final stages, the region contracts to the size of a planetary system; this is a huge reduction in size, by a factor of about one thousand. Since angular momentum is conserved, the very slight rotation that the cloud has at the

(*continued*)

> **Box 9.1** (*continued*)
>
> start of the collapse increases dramatically when the collapse takes place, forming (as we have described in Section 9.1) a swirling disc. Without that slow initial rotation in the cloud, no planets would have formed. In a rotation, there is always a centrifugal acceleration that points radially away from the centre of motion, and without a counteracting force the body will fly away. In a planetary system, that force is, of course, gravity.
>
> Since all the planets in a planetary system formed from the same rotating disc, they all rotate in the same direction. In the Solar System, all the planets move around the Sun and the moons orbit their planets in a counterclockwise motion (as seen from above the system). The rotation of the planets themselves also follows the same motion. That's also because of the conservation of angular momentum.

While the Sun – with 99.8% of the mass of the Solar System – absorbed the vast majority of the material in this disc, the planets have more than 90% of the angular momentum. The young Sun would have had a very strong magnetic field, whose lines of force reached out into the disc. These field lines, linked to charged particles in the gas, acted like anchors, slowing down the rotation of the Sun (so-called *magnetic braking*), while amplifying the rotation of the disc gas that would eventually turn into the planets. The disc probably survived for around 100 million years. Although this may appear to be sufficient time for the planets to form, it may not be so in astronomical terms. As the Sun's light warmed the disc, gas evaporated quickly giving the newborn planets and moons only a short amount of time to accumulate this material. Nevertheless, surrounding the Sun is a complex system of worlds with a wide range of conditions: eight major planets, many dwarf planets, hundreds of moons, and countless smaller objects (see Box 2.2 and Table 9.1).

The initial stages of star formation occur at rather low temperatures, and it is in these early stages that interstellar molecules and dust play important roles, while their influence is limited when the temperature in the star-forming region begins to rise significantly. However, it is the low-temperature stages that are critical to success or failure in the birth of a star. As we shall see, dust grains have specific roles to play in the process, and it turns out that they are an essential component of the whole process in the nearby Universe. It has not always been like that. Star formation in the early Universe was different from that in the local Universe. The major difference between current

Table 9.1 Mass of objects in the Solar System.

Object	Percentage of the solar system mass
Sun	99.8
Jupiter	0.1
Comets (estimate)	0.0005–0.03
The other planets (including dwarf planets)	0.04
Moons and rings	0.00005
Asteroids	0.000002
Dust	0.0000001

star formation and formation of the first stars in the very distant (and very early) Universe is the current presence of dust. As we've seen in Chapters 3 and 4, dust is composed of heavy elements such as carbon, oxygen, and iron, all of which are made in supernovae that occur at the end of some stars' lives. Stars without any heavy elements have never been observed, which most likely means that the first stars were sufficiently massive that they are long dead. The emergence of dust in the cosmic scene following the spectacular end of the first generation of stars, completely changed the fate of the Universe. Without dust, galaxies could not have evolved as they have, stars would not have their present forms, and planets could not exist.

In fact, planetary formation is another consequence (and an important one!) of the presence of dust. The formation of a star is almost always accompanied by the formation of a planetary system. Within the solar proto-planetary disc, moving dust grains and ices embedded in the gas, occasionally collided and clumped together, in a complex and, apparently random way. Through this process, called *accretion*, these microscopic particles formed aggregate bodies, that, growing larger, attracted more matter from the disc and gradually built, in some way, kilometre-sized bodies called *planetesimals*. Planetesimals located in the inner, hotter part of the solar nebula were composed mostly of silicates and metals, while in the outer cooler regions of the disc, ices were the dominant component. The further growth of planetesimals from a few kilometres to a few hundred kilometres across – the nascent planets – produced bodies massive enough that their gravity influenced each other's motions, increasing the frequency of collisions, and the growth rate. We are beginning to understand that this was a violent process, in which the growing planets themselves could have been subjected to catastrophic destructive events. We could witness a faint representation of such events during the spectacular end experienced by the comet Shoemaker-Levy 9 that struck Jupiter in 1994. This was the first collision of two Solar System bodies ever observed by astronomers. The collision produced scars that were visible from Earth.

If the dust is the engine of the process, at the same time the formation of a planetary system is a filter that severely modifies the composition of interstellar dust; those materials passing through the filter were the most robust to destruction and erosion. Just as stardust is modified in the interstellar medium, and becomes interstellar dust, so interstellar dust is modified when it is incorporated in the gas that forms a planetary system.

9.2   AN OVERALL PICTURE OF STAR FORMATION IN THE MILKY WAY GALAXY

Extensive surveys of interstellar regions of the Milky Way in which stars are forming reveals a vast and intricate network of filamentary structures and dark bubbles, interspersed by bright hotspots where new stars come to life. Filaments are nearly everywhere in the interstellar medium, giving rise to a universal web-like structure permeating the entire galaxy.

There is a link between star formation and the filamentary structures in the interstellar medium. In the densest strands, the gas that constitutes the filaments becomes unstable and forms clumps of material bound together by gravity. If dense enough, these collapsed blobs of gas eventually go on to become newborn stars. The filaments come in all sizes, from small ones, only a few light years long, to giant threads extending over hundreds of light years. These filaments all seem to have roughly the same width – about a third of a light-year – regardless of their length. This suggests a common physical mechanism in their origin, possibly linked to the turbulent nature of the galactic gas (interstellar turbulence is discussed in Box 9.2). In fact, such width corresponds to the typical scale at which the interstellar gas undergoes the transition from super-sonic to sub-sonic regimes.

Box 9.2   Interstellar turbulence

A fluid is defined as any substance that flows or deforms under applied shear stress. Fluids should be simple to describe as they seem to be quite ordinary things. After all, it's easy to imagine liquid flowing steadily through a pipe as a uniform column. Regrettably, it is anything but simple!

The peaceful flow that we imagined before is called *laminar flow*, and it occurs when the fluid flows in infinitesimal parallel layers moving at the same pace with no disruption between them [see Figure 9.1 (left)]. The lines shown in the figure are the paths followed by small volumes of fluids. These are called *streamlines*. This ideal laminar flow is most efficient from a pure energy loss point of view; however, it cannot realistically be maintained. The primary disturbance is friction. Even in a straight pipe with a smooth interior, the liquid closest to the wall moves the slowest because it is rubbing against the pipe wall, and tends to slow down the nearest layer. Thus any layer is slowed by its outer layer, so that the liquid at the centre of the pipe moves the fastest [Figure 9.1 (centre)]. We call this property of real fluids *viscosity*. In general, viscosity tends to "stabilize" the flow of a fluid.

Streamlines are smooth and continuous when the flow is laminar, but become messy when any obstruction or other effect creates transversal motion of the fluid particles. The smooth fluid begins to split into *eddies* and *vortices*, which in turn break into smaller swirls, with those swirls "dissipating" into ever-smaller whorls, an unpredictable cascade that degrades the energy from the original smooth stream. Eventually this process stops when the emerging structures have sizes small enough that the particles involved are the atoms and molecules composing the fluid. Their interaction is such that dissipation of energy finally takes place. All these structures affect one another, making the motion of the fluid erratic (one could say *chaotic*), and impossible to predict precisely. In contrast to laminar flow, the fluid does not flow in parallel layers; the lateral mixing is very high, and there is disruption between the layers. We say that now the fluid is *turbulent* [Figure 9.1 (right)]. Turbulence is thus characterized by apparent randomness, with the speed of the fluid at any point continuously undergoing changes in both magnitude and direction.

The gas in galaxies is typically seen to be moving at very rapid, even supersonic velocities, providing clear evidence that the medium is highly turbulent. Despite its importance and ubiquity, however, turbulence is poorly understood. Even its origin is far from clear. Complicating the picture is the

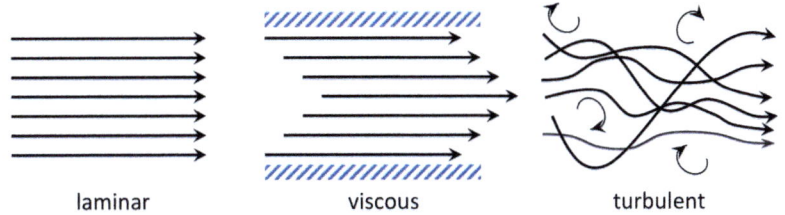

laminar   viscous   turbulent

Figure 9.1   The differences between laminar, viscous, and turbulent flows in fluids.

(*continued*)

Box 9.2   (*continued*)

presence of powerful interstellar magnetic fields that influence the motion of both interstellar dust and gas. This unfortunately makes it hard to characterize the motion of the gas using astronomical observations (Figure 9.1). There could be many reasons to explain why interstellar turbulence is present. Some astronomers argue that turbulence results from star formation itself, as new stars and their associated supernovae drive winds that stir up the interstellar medium. Others point out that gravity alone may induce super-sonic motion in gas as it moves through and across a rotating galaxy. We simply don't know for sure. What we do know is that turbulence is one of the key factors in the energetics of the Universe, and one of the major agents affecting most of the processes in astrophysics.

The accumulation of observational evidence has led to a new scenario to explain how stars of low mass, like our Sun, are born. It is a two-step scheme: first large-scale super-sonic flows compress the gas, giving rise to the web-like filamentary structure; next, gravity takes over the control of denser structures that start to contract. While this collapse occurs, the cloud can fragment into smaller masses, with each fragment giving rise to *pre-stellar cores*, the seeds of future stars, and ultimately to *proto-stars*, stars that are still accumulating mass from their parent clouds. However, filaments constitute only a small fraction of the total mass that makes up the galaxy's interstellar medium, and only the densest of them partake in the highly inefficient process of star formation.

Massive stars exceed the mass of the Sun by several times. Being extremely huge, these stars "live fast and die young", and make significant impacts on their environments. The existence of massive stars presents a challenge to our understanding of star formation, because the enormous radiation pressure that arises as they form, may overcome gravity to prevent the accretion process entirely. Perhaps the best example of a high-mass star is Eta Carinae (see Figure 9.2). This extremely massive object is located about 8000 light years from Earth, and is part of the constellation Carina in the Southern sky. The estimated mass is between 100 and 150 solar masses. The star is probably fewer than 3 million years old, and it's believed that it has fewer than 100 000 years left to live. Because of their masses and energy outputs, these stars must come to life in conditions that are quite different from those found in the birthplaces of their lower mass counterparts.

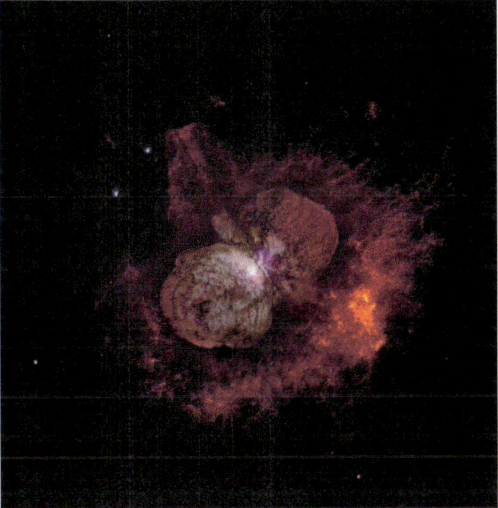

Figure 9.2 The nebula surrounding the star system Eta Carinae containing at least two extremely massive stars. Eta Carinae is variable and very bright, being millions of times brighter than the Sun. Winds from Eta Carinae have blown the two huge bubbles in the interstellar gas that we can see in this image (credit: NASA, ESA, and the Hubble SM4 ERO Team).

Massive stars are observed to form near-gigantic, massive, high-density filaments and in the spherical knots that may arise at the intersection of ordinary filaments. These regions have enormous reservoirs of gas and dust, and may provide enough "fuel" to support the growth of huge stellar embryos. Here the star-formation rate may be very high, eventually giving rise to stellar clusters hosting primarily massive stars. Most stars are in fact formed in stellar clusters, which are groups of stars that are sufficiently close to each other to feel the effects of their neighbouring stars. There may be thousands of stars in a cluster, and in some cases over a million stars. Depending on the stellar density, a cluster can be held together tightly by the effects of mutual gravity, or it can dissolve as the parent gas cloud dissipates or when it intercepts another molecular cloud.

Although the scales over which high- and low-mass stars form are very different, the mechanisms driving their formation appear to belong to a common framework, in which they are parts of a continuous process taking place on all scales: the interstellar material is stirred up, compressed and shaped into a filamentary structure, whose later collapse under gravity and subsequent fragmentation gives rise to a multiplicity of different stars.

If we ignore for a moment all processes inhibiting collapse, and imagine the collapse occurring solely under gravity without resistance, the process takes place in the time, called the *free-fall time*, required for all mass to fall to the centre of gravitational attraction. This is the minimum time necessary for the collapse to occur. Of course, a real collapse is much more complex than this simple description. Parcels of gas at the edge of the cloud and within it begin to fall towards the centre of the cloud, but in doing so they collide with other cloud material, and this friction generates heat. Evidently, the gravitationally collapsing core must be able to rid itself of the heat generated during the infall, otherwise the gas would heat up and consequently the increased pressure would terminate the process. The amount of heat to be dissipated depends on the nature of the various supporting pressures (*e.g.*, thermal and radiation pressures). Generally, the only effective cooling process in interstellar conditions is the emission of infrared, sub-millimetre, and radio waves. Any specific cooling mechanisms must be able to operate at the temperatures of cold dense cores, *i.e.*, around ten degrees. There are two mechanisms that can do this: emission from dust and emission from molecules. We shall discuss them in Section 9.6.

If the collapse of the cloud is able to continue, and the cloud is not uniform (as is always the case), then the higher densities achieved locally make it possible for individual parts of the cloud to become unstable, leading to the fragmentation of the original cloud into an association of smaller, denser pockets of gas. The collapse and further fragmentation of the cloud can continue until gas densities become so high that the gravitational potential energy released as heat cannot be radiated away. Associations of dense regions within a Milky Way molecular cloud can be observed using telescope arrays (such the ones described in Box 6.3) to obtain sufficient angular resolution. Even with the help of these powerful instruments, it is difficult to disentangle single regions in the astronomical images because of their crowding along the line of sight.

At this stage, central temperatures are able to rise significantly without affecting the stability of the structure, if gravity is strong enough. The cloud fragment has become a roughly spherical core with radius on the order of 10 000 times the Earth-to-Sun distance (*i.e.*, 10 000 AU; see Table 1.2 for a definition of AU), with gas infalling from all directions. After 100 000 years, the rotation has caused the development of a circum-nuclear disc, and an outflow of gas is produced along the symmetry axis of the disc. The outflow begins to open its way through the infalling material, and after one million years the rate of infall decreases significantly. A proto-star with a well-established outflow, a weaker infall and a disc, is born (see Figure 9.3). Eventually, the accretion ceases, the outflow weakens and the disc is largely eroded, although a remnant disc remains in orbit around the star. The disc – what we may now call *a proto-planetary disc* – may be around 100 AU in diameter at this stage.

Our Solar System is embedded in the Öpik–Oort Cloud, a shell of icy planetesimals that exists in the outermost reaches of the Solar System (see Box 2.2), a relic of the part of the original cloud that generated our planetary system. The Öpik–Oort Cloud is

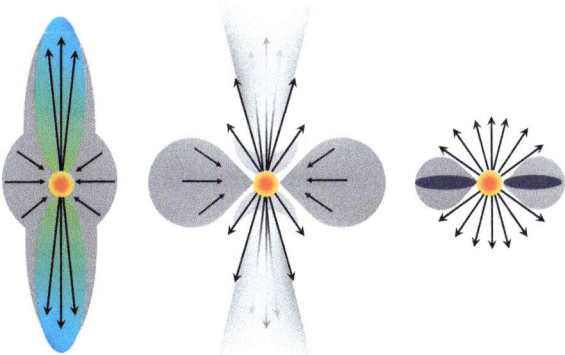

Figure 9.3   How infall and outflow of gas occur simultaneously during the formation of a star. In the left-hand image, rotation causes the formation of a disc around the star. Gas and dust fall inwards to the star from the disc while outflow occurs along the rotation axis. In the central and right-hand images, the outflow gradually widens and reduces the infall from the disc. The infall is eventually suppressed.

roughly spherical, and it is supposed to extend a quarter of the way to the closest stars (known as the Alpha Centauri system). The planetesimals that are populating this region may have originated much closer to the Sun, but have been ejected by interactions with newly-formed planets. The equivalent may also be found in other planetary systems.

This rough description applies to the formation of stars of low and moderate mass (up to around eight solar masses). However, it does not apply accurately to stars of greater mass, which form in much more massive ridges and hubs in the filament network. In particular, the evolutionary timescale is much shorter.

However, not all molecular clouds end up forming new stars, and not all the mass in a star-forming cloud goes into stars. What determines when and where stars form, and what determines their masses? One aspect of particular importance is knowing how many stars of each size there are, that is, knowing the initial distribution of stellar masses in a newly formed stellar system, the so-called *initial mass function*. Although non-universal, this quantity appears to be similar for almost all the stellar systems that have been observed. In general, the larger a star, the shorter its life. The great interest in the initial mass function lies in the fact that, if it were well understood, it would shed light on how the stars formed, and on how the conditions under which stars form determine their future, and that of a galaxy. Our Sun, a typical star, will live for about 10 billion years, while the least massive stars, with about a tenth the mass of the Sun, live 100 times longer. By contrast, the most massive stars live for only a few million years. The observed initial mass function has relatively few massive stars, while Sun-sized stars are comparatively abundant. Stars somewhat smaller than the Sun are even more common, but then stars of decreasing mass, down to one-tenth of the Sun's mass or even less, decrease in numbers. The precise statistics for low-mass stars are somewhat uncertain because they are faint and hard to detect.

As we have seen above, during the collapse, a cloud can fragment into smaller masses, the birthplace of stars. Evidently, these pieces cannot have the same size, and their mass distribution appears naturally linked to the initial mass spectrum of the nascent stars. This, however, cannot be the whole story since other factors may play important roles, such as chemical composition. In fact, dust and metals, magnetic fields, and dynamical interactions all determine the distribution of initial stellar masses. Also, it's often difficult to define a core observationally, as these regions have a tendency to blend together, emphasizing the vagueness of their characterization.

9.3 FORMATION OF PLANETS AND COMETS IN THE SOLAR NEBULA

At the end of the collapse phase, the solar nebula was at its hottest. Without gravitational energy to heat it, most of the nebula started to cool, although at the centre, the newly-formed Sun maintained high temperatures in its immediate neighbourhood. Turbulent motions and magnetic fields were lessening the angular momentum, slowing down the rotation of the disc, and the disc gradually stabilized. The temperature within the disc decreased with increasing distance from the Sun, allowing the gases to interact chemically to produce new compounds. The process was similar to dust formation in the expanding envelopes of cool stars, or the way raindrops on Earth condense from moist air as it rises over a mountain. In the inner Solar System, it was too warm for volatile materials such as water and ammonia to condense. At the same time, the solar wind removed volatiles from the inner part of the disc, driving them towards the external, much colder regions. When volatiles crossed an invisible boundary called the *snow line*, where ices can exist without being melted or destroyed by the Sun's heat, they condensed and were incorporated into the nascent giant planets. Today, this snow line is slightly beyond the orbit of Mars. A more typical snow line around a young star like the Sun would have been around three times the Earth-to-Sun distance, in the region between the current orbits of Mars and Jupiter.

Figure 9.4 shows the chemical condensation sequence in the solar nebula. The first materials to form solids were metals, then silicate-forming rocks. As the temperature fell, water-rich silicates and sulfur compounds followed. Further from the Sun, cooler temperatures allowed oxygen to combine with

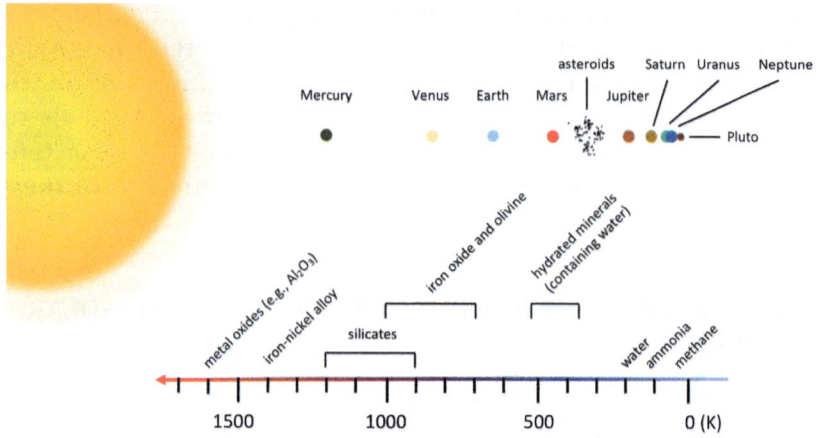

Figure 9.4   Chemical condensation in the solar nebula. The temperature scale shows how the temperatures of solid bodies heated by the Sun vary with distance from the Sun. The condensation temperatures of some substances are also marked on this scale. The positions of the planets are located appropriately in the diagram, allowing inferences to be drawn about the chemical nature of planets at different distances from the Sun.

hydrogen and condense in the form of water ice, and – further still – carbon and nitrogen, as they crossed their own snow lines combined with hydrogen to make ices such as methane and ammonia.

Figure 9.4 also reports the positions of the planets and other constituents of the planetary system. This sequence of events explains the basic chemical composition differences among various regions of the Solar System. For instance, water-rich silicates, are now found abundantly among the asteroids. In the inner parts of the disc the temperatures never dropped low enough for ice or carbonaceous organic compounds to condense. Thus, those materials were generally missing, and they are lacking now on the innermost planets.

The composition of the disc was practically the same as the star – mostly gas (hydrogen, helium, and carbon monoxide), with some dust and ices. All these materials started to glue together, meeting *slowly* and sticking together *very gently*. The accretion process was not the same in all parts of the disc: it's much easier to make a ball with snow than with sand. Beyond the snow line, ices and dust very quickly joined into larger and larger chunks, until most of the solid material was in the form of a solid core. When large enough, a core started to develop a significant force of gravity, and then interacted with other cores to build even bigger cores. These proto-planets grew and became large, with masses ten times greater than Earth. These proto-planets of the outer Solar System were so massive that they were able to gravitationally attract and hold the surrounding gas. Planets like Jupiter or Saturn therefore formed very rapidly. If their formation had been slower, the gas would just have vanished – because the whole process was in competition with solar wind and radiation from the young Sun that were blowing away the remaining supply of lighter gases. It took about 20 million years for the gas to be removed. This is the reason why the compositions of Jupiter and Saturn are closer to that of the Sun than those of other planets. Uranus and Neptune captured much less gas. In fact, these two planets have compositions dominated by the icy and rocky building blocks that made up their large cores rather than by hydrogen and helium. To a large extent, this is also what happened in the inner Solar System where disc materials were closer to the star. There was no ice, so building up a planet was a slow process. Therefore, the inner planets didn't grow big enough to accrete the gas.

While the intermediate steps are not well understood, ultimately several dozen centres of accretion seem to have grown in the inner Solar System. Some of these centres attracted surrounding planetesimals until the centres acquired a mass similar to that of Mercury or Mars. At the end of the formation epoch, the inner Solar System was populated by about one hundred planetary embryos of size between that of the Moon and Mars. These embryos kept colliding and merging for a little less than 100 million years. Such continuous mutual interaction went on until the four terrestrial planets we know today – Mercury, Venus, Earth, and Mars – took shape. One of those collisions is supposed to have formed the Moon, while another one might have removed the outer envelope of the young Mercury. The colliding planetesimals, being accelerated by the gravity of the proto-planet, struck with enough energy to melt or vaporize both the projectile and a part of the impact area. Multiple collisions were able to heat an entire proto-planet above the melting temperature of rocks. This gave rise to planetary differentiation, with denser metals, such as iron, sinking gravitationally to the centre of the body and lighter silicates rising toward the surface. In the process, the inner proto-planets lost part of their content of lighter gases, leaving more of the heavier elements and compounds behind. The final result is that the material of these planets evolved into the separate components of *core*, *mantle*, and *crust*. The newly formed planet also contained relatively small quantities of radioactive elements. As these radioactive atoms spontaneously disintegrated by nuclear fission, energy was released in the form of heat, in addition to the energy generated by accretion. While accretion deposited heat billions of years ago, radioactive decay remains a source of heat in terrestrial planets, although it was stronger in the past.

After the Sun and its planets and their moons were formed, there were many planetesimals and other debris ranging from ice and rock balls to tiny dust grains that did not initially accumulate to form the planets. What was their fate? Comets, asteroids, and meteorites are surviving remnants from the processes that formed the Solar System. Planets and moons, as well as the Sun are also, of course, products of these formation processes, although as we have seen, the material in them has undergone a wide range of changes. Detailed analysis of data collected by the *Rosetta* probe – a spacecraft that spent the past 10 years traveling to a comet nearly 280 million miles away – showed that comets are ancient left-overs of early Solar System formation, and not younger fragments resulting from subsequent collisions between other, larger bodies.

The European Space Agency launched the *Rosetta* spacecraft in 2004, and 10 years later it reached its destination, a comet called *67P/Churyumov-Gerasimenko*. In collaboration with NASA, *Rosetta* was put into orbit around the comet in August 2014, and then the robotic lander *Philae* was dropped onto the surface of the comet in the November of the same year, providing an unprecedented close-up examination of a comet. The comet has a barbell-shaped core roughly five kilometres along, of low density, high porosity, and extensive layering, suggesting that the lobes accumulated material over time before they merged. Such an unusually high porosity shows that the comet did not grow through violent collisions, as these would have compacted the fragile material. Thus, the nucleus is composed mainly of non-volatile dust, with only minor contributions by volatiles. For most of their life time since formation, comets have orbited the Sun at large heliocentric distances so that they remained almost unaffected by solar radiation. However, gravitational disturbances by the giant planets can change their orbital parameters over time and, thus, comets can get closer to the Sun.

Comets formed far away from the Sun. They're made of dirty ice, which would melt if they were as close to the Sun as the asteroids. Asteroids formed from small pieces of rock and metal, just like the rest of the inner Solar System. The reason why they did not grow in size into a planet is the strong gravity of Jupiter, which stirred them up, and makes them go so fast that when they run into each other, they usually bounce off or break apart instead of sticking together. Most meteorites are believed to be fragments of asteroids either knocked out of their orbit of the Sun, and into Earth-crossing orbits through collisions with other objects, or through the interaction of gravitational forces exerted by the Sun and Jupiter. Some meteorites come from the Moon and Mars. These crustal rocks were ejected into space by impacts of asteroids or comets with the Moon or Mars of enough power to launch some of the impact-produced debris into Earth-crossing orbits. Occasional accretion of asteroids leads to the formation of larger bodies like Ceres, the only *dwarf planet* in the inner Solar System. When comets penetrate the inner parts of the Solar System, and travel close to the Sun, they melt, and the ice is turned to water vapour forming the long tail that we see in the sky. After a few rendezvous with the Sun, catastrophic melting may break the comet into little pieces; see Figure 9.5.

Planets, asteroids and comets need to find stable orbits. The vast majority of asteroids in the Solar System are found in a region out beyond Mars, called the *Asteroid Belt*, located between the orbits of Mars and Jupiter. In addition, other orbital families exist with significant populations, as gravitational perturbations caused by planets and collisional effects drive a continuous migration that brings asteroids from the main belt closer to the Sun. In addition to the Asteroid Belt, there is another region in which stable orbits are possible, where left over planetesimals can avoid impacting the planets or being

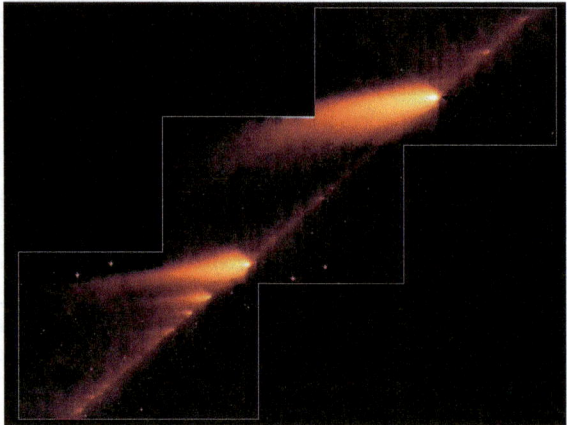

Figure 9.5   The break-up of a comet. Comets are fragile objects, and the heating they receive from close passage to the Sun may cause them to fragment. This false-colour infrared image is of debris from Comet 73P/Schwassmann-Wachmann3 [credit: NASA/JPL-Caltech/W Reach (SSC/Caltech)].

ejected from the system. This is called the *Edgeworth–Kuiper Belt*, and is located beyond Neptune. The planetesimals and their fragments that survive in these special locations are now called asteroids, comets, and trans-Neptunian objects. There is also a region roughly overlapping the Kuiper Belt, the *scattered disc*, composed of objects having highly eccentric orbits with large orbital inclinations. Close encounters with giant planets cause smaller objects to be flung to the outer Solar System. The bodies of the Asteroid Belt cross the orbits of the inner planets, including several thousand large ones that intersect the Earth's orbit (the so-called *Near Earth Objects*, or NEOs). If one of these intruders hit Earth it could devastate life on our planet. An NEO called Apophis with diameter 370 m is predicted to have a close encounter with Earth in 2029; Apophis returns again for a close encounter in 2036, but there is a negligible possibility of impact on Earth on that occasion. An extra-terrestrial impact is one of the hypotheses that may explain the mass extinction that marks the end of the Cretaceous Period of the geologic times-cale, when all dinosaurs were wiped out. The impact choked the sky with debris, reducing the amount of light that reached the Earth's surface. So, it had an impact on plant growth, throwing a spanner in the works of photosynthesis and messing up the food chain. Once the dust settled, greenhouse gases locked in the atmosphere would have caused the temperature to soar, and stay about 5 °C warmer over a period of something like half a million years.

The four largest asteroids in the belt are Ceres, Vesta, Pallas, and Hygiea. They contain half the mass of the entire belt. The rest of the mass is contained in countless smaller bodies (see Box 2.2). Collecting all the material that exists today in the Asteroid Belt, would make a tiny world smaller than Earth's moon. This means that if, as long supposed, a fifth rocky planet had ever formed there, it would have been rather insignificant! The location of the asteroids between the gas giants and terrestrial planets suggests that they represent the boundary between two distinct reservoirs of planet-forming material: relatively anhydrous bodies from within the snow line and more water-rich bodies from outside the snow line.

9.4   PLANETARY MIGRATION AND EXOPLANETS

The collapse of the solar nebula, the formation of the disc, and – within the disc – the accretion of materials explain why orbiting bodies are so sparse and where we can find them, including the various types of planets. These ideas also explain why the planets all lie in about the same plane, and orbit the Sun in the same direction (see Box 9.1). Despite their diversity, all the objects in the Solar System formed together, along with the Sun, precisely as a system. The basic story does sound ordered, with small rocky planets inside and more massive planets outside, all these planets reasonably forming in the same part of the Solar System they reside in today.

The discovery of exoplanets shook things up, because the first exoplanet that was found near another star was a planet like Jupiter, but very close to its star and very hot, a *hot Jupiter* (as it has been hereinafter called with very little imagination). This is in strict contrast with the previous scenario: how could a gigantic planet attract all the gas if it is so close to the star? There is no ice, it would just melt, vaporize, because the star would strip away that gas faster than the nascent planet's gravity could pull it close. This is exactly what we observe happening around some of the hottest exoplanets. The answer is rather simple: such planets are thought to have formed outside the snow line, and later moved inwards to their current positions. In other words, planets *migrate*.

Looking more closely, our Solar System is much more weird and complex than it seems at first sight. There are huge regions of debris that aren't collected into any one object: the Edgeworth–Kuiper Belt and the Öpik–Oort Cloud. Why is there an asteroid belt in the middle of the Solar System? Having a bigger orbital path than Earth, Mars seems too small, while the Moon came out from a monstrous object (perhaps Mars-sized) that struck the Earth in its early history. Indeed, Nature starts most chains of events randomly. Thus, it is very possible that the Solar System planets formed in orbits that were not stable over their lifetimes. Occasionally that instability caused the planets to impact onto each other. More frequently, they approached very closely so that their orbits were deeply modified. This "slingshot effect" is frequently used with spacecraft. For example, for the first time in history, a human-made object, *Voyager 1*, entered interstellar space, crossing the edge of the Solar System delineated by the Öpik–Oort Cloud on August 25, 2012. The spacecraft gained the energy to escape the Sun's gravity completely by performing slingshot manoeuvres around Jupiter and Saturn.

If the Voyager flight is the result of careful calculations and planning, in space, encounters occur randomly, resulting in random slingshots. During such events, planets become unbound from their systems, and drift free. One example could be the huge object that collided with Earth to form our Moon, and now is nowhere to be found. A few of these *rogue planets*, unattached to any identifiable star, have actually been observed, but there could be 50 billion rogue planets adrift in the Milky Way (see discussion in Box 2.2).

Recent numerical simulations by Konstantin Batygin and Gregory Laughlin show that inward Jupiter migration may have scrambled an early generation of inner planets, before it returned to its current orbit. Scattered left overs of gas in the disc, denser in some areas than others, are thought to have caused a loss of angular momentum in Jupiter during its cruise around the Sun. As Jupiter migrated in toward the Sun, it might have gathered material that would have otherwise belonged to Mars, so that planet ended up being smaller than

Earth. In this scenario, the original population of the Solar System inner planets featured a large number of larger-than-Earth planets (the so-called *super-Earths*), which are thought to have had substantial gaseous atmospheres. Jupiter migration could have set off a series of collisions that smashed these worlds into pieces. As a consequence, the four terrestrial planets that are currently orbiting the Sun must have formed hundreds of millions of years after the Sun and the rest of the Solar System. The subsequent formation of Saturn caused a gravitational interaction that drew Jupiter back out again to its current orbit. The amount of material left behind from this catastrophic journey was much less than the original content of the disc, and this reduced amount of material was the source from which the four small terrestrial worlds, Mercury, Venus, Earth, and Mars, arose. These planets possess very thin atmospheres compared to those seen in other planetary systems. On its way back to the outer regions of the nascent planetary system, Jupiter's gravitational pull attracted asteroids that had been formed beyond the snow line inwards towards what is now the Asteroid Belt, and towards the region where Earth was forming. This intense period of planetesimal bombardment, occurred late in Solar System history (the so-called *late heavy bombardment*), and it may have provided enough ice to account for Earth's oceans.

This description – known as the *Grand Tack hypothesis* – is one among the possible stories about the origins of our planetary system, and others may be found (and actually are). Nevertheless, systems like our own Solar System seem to be a minority, and 'truly' Earth-like planets, with Earth masses, hard surfaces, and small atmospheres are not very common at all. For centuries, planets beyond our Solar System have been matter of speculation. Planets are billions of times fainter than their parent stars, so they were thought to be unobservable. In the last two decades, however, astronomers successfully developed indirect detection methods, based on the effect that planets have on the luminosity of their parent star. Currently, the *Extrasolar Planets Encyclopedia* lists more than 4000 exoplanets, some of them seemingly coming directly out of science fiction movies. These findings reveal again that our Solar System is very unusual. Typically, a planetary system is composed of a few super-Earths (originally this might be the case for our own system) – rocky planets up to 10 times the mass of Earth – orbiting much closer to their stars than Mercury does to the Sun. These super-Earths are usually not only rich in rock, but also in volatile materials that vaporize when heated.

Astronomers classify the various types of exoplanets as gas giants, hot Jupiters, super-Earths, water worlds, and exo-Earths, which can be divided in three groups, terrestrials, gas giants, and mid-sized gas dwarfs. There are two clear dividing lines: one between 1.5 to 2 times the radius of the Earth, and the other at 4 times larger than Earth, which appear to mark changes in composition. Smaller planets are likely to be rocky, while larger ones are probably gas giants. In the middle, the rocky cores of gas dwarfs formed early enough to accrete some gas, although they were unable to grow as large as gas giants like Jupiter. We show in Figure 9.6 the currently confirmed exoplanets, classified into eighteen thermal-mass categories collected together in the so-called *"Periodic Table" of the Exoplanets*,

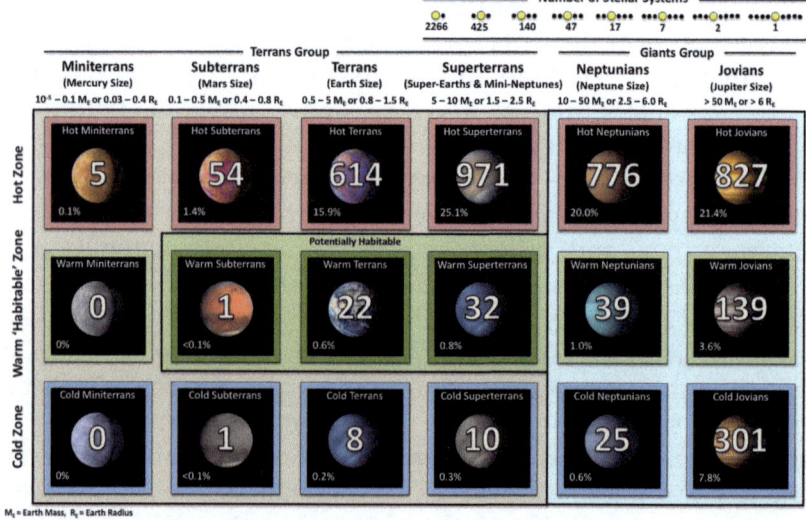

Figure 9.6   The "Periodic Table" of the Exoplanets. Just as the periodic table of the elements arranges those elements into different chemical categories, so this 'periodic table' arranges the known exoplanets into various categories defined by planet temperature and size. The numbers of detected exoplanets are shown in each category (credit: PHL@UPR Arecibo).

dividing most of the known exoplanets into six mass/size and three temperatures groups. *Miniterrans* are low-mass planets similar to Mercury and the Moon. *Subterrans* are comparable to Mars, *Terrans* to Earth and Venus, and *Superterrans* are a transition group between terrans and Neptunians. *Neptunians* are similar in mass to Neptune and Uranus, and *Jovians* to Jupiter and Saturn, or larger.

9.5   HOW TO BUILD A PLANET FROM SCRATCH

Before the formation of planets there was the star-disc system. Dust grains in the disc somehow clump together and eventually grow into an entire world. But exactly how do dust particles stick together? Dust particles are attracted to each other by electrostatic forces, building pebble-size clumps, similar to the "dust bunnies" that may accumulate under the bed or in corners of a room. However, planets cannot form directly through the accumulation of bigger and bigger objects because eventually the impacts would dislodge previously attached aggregates, stalling growth. For sizes in between a millimetre and hundreds of kilometres, dust grains don't stick. Something else must be guiding the early growth of planets.

It has generally been assumed that tiny dust grains orbiting in a proto-planetary disc are stable, moving along with the gas, or settling out of the gas, if they are big enough – just as soot can settle out from a fire. However, the interaction between gas and dust actually appears to be rather different. Gas streams around a grain, like water around a rock, with the same thing happening to another dust grain nearby. If the grains are close enough, the two flows might interact, and if there are many grains gathering together, as in the case of planet formation, this would result in a gentle channelling of dust grains into clumps, allowing the new-born aggregates to rapidly grow in size. Such a process, called *streaming instability*, concentrates dust into compact gatherings and filaments of grains, whose densities may be hundreds of times larger than those typical of the disc. With such high densities and reduced relative velocities, grains may then coagulate due to self-gravity, forming the seeds around which planetesimals can grow. The engine of the whole process is the drag – the same force that makes it harder to drive against a headwind – that slows down dust, allowing other particles to pile up behind it. Then, the aggregation process can proceed through gravitational and not dynamical interaction, forming structures across a wide variety of scales. The process keeps working and produces large chunks that attach together in kilometre-sized objects. These solids are now of sufficiently respectable sizes to be called planetesimals. The process is supposed to continue merging solids as big as a few tens of kilometres up to a few hundred kilometres, comparable to the size of the dwarf planet Ceres. At that stage, the drag no longer affects planetesimals, and the game gets tough, being essentially driven by the motion of the nascent proto-planets.

Planetesimal formation through streaming instability is interesting in the context of comet formation, because it predicts highly porous, low-density objects with very low tensile strength. The unplanned landing location of the lander Philae turned out to be a lucky event, because the site was cleaned of cometary dust, leaving behind the inner 'true' comet surface, including plenty of cracks and granular components. The images reported the very building blocks of the comet, millimetre-to-centimetre-sized aggregates reminiscent of pebbles on a beach. This is very consistent with the idea that comets emerge by gentle gravitational aggregation of materials concentrated through streaming instability.

Streaming instability belongs to a broader class of phenomena called *resonant drag instability*: when particles move through gas, a version of this instability appears. These phenomena might be occurring all over the Universe: from how particular stars age and die by blowing off gusts of stellar wind, to how volcanic ash settles in a planet's atmosphere. The size of dust ejected during volcanic activity is important for planetary climate: the more that ash grains clump together as they fall, the less sunlight is extinguished, reducing the dust's cooling effect on the planet. Such kinds of dust–gas interactions could also be happening here on Earth during volcanic eruption or in the dynamics of stratospheric aerosols, so that drag instabilities could be important for our understanding of climate change.

9.6   THE ROLES OF DUST IN STELLAR FORMATION

As we have seen, star formation begins in cold, dark, and dense molecular clouds. It does not occur in the general interstellar medium, where radiation from hot stars pours large amounts of energy into the gas. This implies that potential star-forming regions must be shielded from the interstellar radiation field. Evidently, this is the first important role of dust in star-forming regions.

In addition, dust plays other important roles in the process of star formation: (i) infrared emission from dust removes the gravitational energy of collapsing clouds, allowing star formation to take place; (ii) dust grains may, in very dense regions, become significant carriers of negative charge, and thereby may influence the dynamics of collapse; (iii) molecules such as carbon monoxide in cloud interiors are also protected from photodissociation by ultraviolet starlight and have a long life time, allowing a complex chemistry to build up – the molecules formed in this chemistry are important because they allow astronomers to trace the presence of gas through molecular emissions in the millimetre and sub-millimetre wavebands, which are unaffected by dust extinction, thus represent a loss of energy to the gas, in fact a cooling mechanism; and (iv) infrared emission from dust provides an effective probe for the star-formation processes.

Indeed, cosmic dust plays a crucial role in the appearance of the Universe, re-processing fully half of all non-primordial radiation. In the early 1990s, the Cosmic Background Explorer satellite measured the absolute energy spectrum of the Universe in the far-infrared and sub-millimetre region. These measurements, along with previous observations of nearby galaxies in the 1980s, showed that the Universe emits an energy density comparable to that in the then more traditionally studied optical and UV spectral ranges. About 99% of the energy released by galaxies in the far-infrared and sub-millimetre wavebands is produced by thermal emission from dust grains. Since dust heating and emission is mainly due to local star formation, the implications were remarkable: by ignoring the far-infrared and sub-millimetre emissions, we were missing roughly half of the star-formation activity in the Universe.

9.6.1 Cooling in Contracting Cores

In a typical spiral galaxy, about one-third of the energy radiated by stars is absorbed by dust and re-emitted at long wavelengths. Evidently, radiation from dust grains can be an important coolant for a dense core that absorbs all the external starlight that falls upon it, or which generates heat from gravitational collapse. The spectral energy distribution of the re-emitted radiation spectrum is particularly sensitive to the size distribution of the dust grains. While large grains have low temperatures that vary little, small grains – say, less than about 0.01 µm in radius – have temperatures that fluctuate widely (up to around 50 K), because the arrival of a single ultraviolet photon is sufficient to cause the grain temperature to sharply rise to very high values.

Simple and relatively abundant gas-phase interstellar molecules such as CO and OH can be effective radiators at the temperatures of cold, dark interstellar clouds, through emission in their rotational spectra. During the collapse, as the density increases, the rate at which molecules collide with each other also increases. In the process, a collider (*e.g.*, the abundant hydrogen molecule) knocks a bound electron in *e.g.*, a CO molecule to an *excited state*. The excited state describes an atom, ion or molecule with an electron in a higher than normal energy level than its ground state. The return to a lower energy state is called *decay*, during which generally a photon is emitted. The net effect is to convert thermal motion into radiation that freely escapes from the region, subtracting energy from the contracting cloud. The process works more efficiently when the thermal energy is close to the difference in energy between the ground and excited states (see Figure 9.7). For example, the molecule CO (the second most abundant molecule in space) has the first excited state approximately 5 K above its ground state. This state can readily be collisionally populated in gas with temperatures as low as ~10 K and will radiate as the populations decay. Molecular hydrogen (the most abundant interstellar species) is useless as a coolant at these low temperatures, as the molecule has an excited state at 512 K above its ground state, and can only

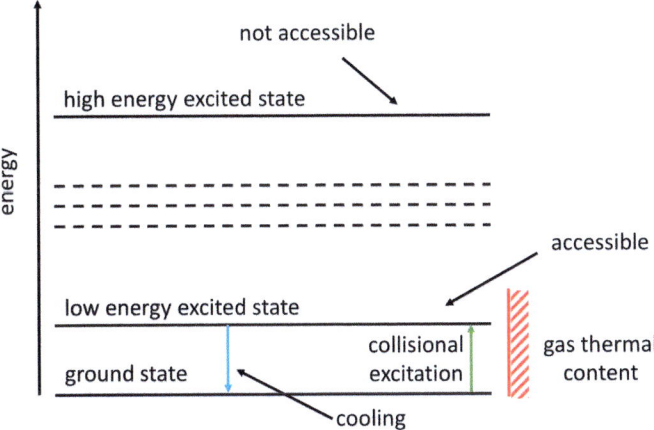

Figure 9.7 Molecular energy levels. Collisions in the interstellar gas may have enough energy to raise a molecule from its ground state to an excited state. The molecule may then radiate this energy away and return to the ground state. In this process, energy has been taken from the gas and radiated away, so cooling of the gas has occurred. However, the energy in the gas (represented in red) is usually only enough to reach low-lying energy levels of the molecule, if they exist. Higher energy levels of the molecule are often not accessible.

be excited in much warmer gas. However, this molecule, H_2, is believed to have played an important role in the formation of the very early stars.

The chemistry of dense interstellar gas has been extensively studied over many years (see Chapter 6), and the ability of the more abundant molecular species to cool the gas through emission in rotational transitions has been well established. Thus, it is rather straightforward to determine the overall cooling rate for a dense interstellar core *via* such processes. Unfortunately, the abundances of the coolant molecules are not well determined. As we described in Chapter 8, molecules tend to freeze-out on the surfaces of dust grains, forming molecular ices, and this process will certainly occur in cold, dense cores where the freeze-out will be rapid at high densities. In fact, the freeze-out time is short compared to the free-fall time (see Section 9.2). In some circumstances, the freeze-out time may be so short that the usual molecular line tracers of cold, dense gas could be entirely missing. However, a complete absence of emission from gaseous molecular tracers is unusual, and dense cores usually show both solid-state features due to ices and spectral lines due to gaseous molecules, in varying proportions. Thus, non-thermal desorption processes can be effective in maintaining a population of gaseous molecules that may act as coolants of dense cores.

9.6.2 Non-thermal Release of Molecules Frozen on Dust Surfaces

Species on or near the surface of a dust grain can be desorbed (removed from the surface) by photons of the interstellar radiation field, if the grains are not too heavily shielded by surrounding dusty gas. Photodesorption by ultraviolet radiation of water molecules may be efficient, because H_2O molecules are very readily photodissociated into H + OH, where ultraviolet radiation is intense, with both fragments having excess energy. They may leave the ice in about 1% of these events, effectively desorbing one H_2O molecule per 100 ultraviolet absorptions. However, sometimes the extinction can be so huge that the cloud gets really dark, and photodesorption driven by starlight is most unlikely.

However, the Universe is permeated by cosmic rays. Cosmic rays penetrate almost everywhere in the interstellar medium, whereas ultraviolet and visible radiation is excluded from denser dusty regions. These are fast-moving atomic nuclei (mainly hydrogen nuclei) and electrons with energies ranging up to ~10^{20} eV. Cosmic rays with energies of a few millions of eV are very much more numerous and much more effective in interacting with cold atoms and molecules. Cosmic rays deposit a small amount of heat when they pass through dust grains, causing some local transient thermal desorption. On large grains, the heat is deposited in a cylindrical volume along the track of the cosmic ray, and thermal desorption of weakly bound molecules such as CO can occur from the heated areas of the surface at the intersection of the cylinder and the surface. Where the grains are small enough, the cylinder encompasses the entire grain, so desorption occurs from the entire surface area of small grains. During their journey across a cloud, cosmic rays may populate hydrogen molecules to highly energetic states, whose decay generates a weak ultraviolet field that may drive photodesorption.

9.6.3 Dust and Magnetic Fields

The magnetic field is an essential ingredient in present-day star formation. Observations have shown that molecular cloud cores have magnetic energy comparable to gravitational energy. Many years ago, in the 1960s, Leon Mestel pointed out that a typical region of the interstellar medium could not contract to stellar densities, because it simply contains too much angular momentum. At the same time such a region contains too much magnetic flux to be able to collapse. Thus, the magnetic field has to find a 'compromise' that enables the removal of angular momentum, and the escape of magnetic field from the collapsing material. Where the ionization fraction is reasonably high, the field and the charged particles are essentially locked together. However, if the ionization fraction is sufficiently low, the field and the charged particles may begin to separate in a process known as *ambipolar diffusion*. This term means that the magnetic field in a dense core may begin to drift out of the core, so that support against gravitational collapse is weakened. If ambipolar diffusion is rapid compared to the collapse timescale, then magnetic support against gravity may not be significant. Approximately, the ambipolar diffusion occurs over times of the order of one million years or so (and thus it is of the order of the free-fall time), and scales linearly with the fractional ionization. So, the question of whether magnetic support can play a role depends on the value of the fractional ionization in the core, the higher the ionization content, the longer the timescale. As a consequence, since dust grains provide shielding of molecular regions from starlight, they reduce the ionization levels and speed up the formation of proto-stellar cores.

Dust grains may play a more straightforward role. Because of dust shielding, in the inner part of a collapsing core, the major driver of the ionization is the flow of cosmic rays through the cloud. Cosmic rays ionize molecular hydrogen forming H_2^+ which – in rapid reaction with another H_2 molecule – gives rise to H_3^+. This latter species is an important one in gas-phase chemistry, because of its ready propensity to donate protons to almost any other species. In particular, it can donate a proton to the molecule CO to form HCO^+. If CO is abundant in dense gas, then HCO^+ may be the most abundant ion. The fractional ionization is estimated to be one electron per hundred million hydrogen atoms inside dark clouds. This means that it is the abundance of CO that controls the ionization, and this abundance is determined by CO freezing-out over dust surfaces (whose timescale is lower than both free-fall and ambipolar diffusion times), and by desorption processes.

Recently, the magnetic structure of the region called Pillars of Creation – a region of towering filaments of cosmic dust and gas located at the heart of the Eagle Nebula (see Figure 9.8)

Figure 9.8   The Pillars of Creation. Star formation is occurring in dense gas at the tips of these pillars, but the whole region is being eroded by intense radiation from a very powerful star that is not shown in this frame. Less dense gas between the pillars has been removed by this erosion, leaving the denser star-forming regions and the gas behind them in this prominent (but highly transient) structure [credit: NASA, ESA, and the Hubble Heritage Team (STScI/ AURA)].

made famous by an iconic 1995 Hubble Space Telescope image – has been mapped, showing the existence of extremely subtle magnetic fields. Dust grains within the pillars are aligned with each other due to the magnetic field. These aligned particles emit polarized light, allowing astronomers to trace the direction of the magnetic field at various locations. The field runs along the length of each pillar, perpendicularly to the external magnetic field. This magnetic configuration suggests that the Pillars have evolved thanks to the strength of the magnetic field, so that stars could be formed by the collapse of clumps of gas being slowed down by magnetic fields, and resulting in a pillar-like formation. At the same time, magnetic fields may also be currently slowing the destruction and fragmentation of the region.

CHAPTER 10

Where and How Does Life Begin?

10.1   IS THE ORIGIN OF LIFE LINKED TO COSMIC CHEMISTRY?

The only known examples of life are found on our planet. It's easy to take life for granted, but its existence raises a very difficult question: where did life on Earth come from? There are multiple hypotheses concerning this topic, among them the possibility that life may have spontaneously generated (so-called *abiogenesis*): in the remote past, precursors to life such as amino acids and proteins somehow arose and managed to arrange themselves into self-replicating pre-cellular life forms. We actually don't know that much about how life started on Earth. In fact, we don't even know if life started on Earth, at least, not its constituent parts. What we do know, and have done for many years, is that complex organic molecules forming the building blocks of life are able to exist in space. Thus, understanding the production of organic material in space, in particular at the early stages of star formation, is critical to tracing the evolution from simple molecules to potentially life-bearing chemistry. One of the big questions is whether products of routine cosmic chemistry

Dust in Galaxies
By David A. Williams and Cesare Cecchi-Pestellini
© David A. Williams and Cesare Cecchi-Pestellini 2020
Published by the Royal Society of Chemistry, www.rsc.org

produced through the long cosmic history of the biogenic elements may have contributed to the early Earth's organic pool, facilitating pre-biotic molecular evolution. Alcohols, sugars and amino acids are detected in meteoritic and cometary material, some of these organic species are observed in regions where solar-like stars and planetary systems are currently forming, and even in far-distant galaxies. It is the versatility of carbon in its chemical bonding that drives the complexity of the chemistry in the Universe, making organic chemistry very common, perhaps even the norm in space.

The organic molecules discovered in space until a few years ago consisted of a backbone of carbon atoms arranged in a single and roughly straight chain. An important step forward in the quest for pre-biotic chemistry was taken with the detection of the first chiral molecule, propylene oxide, and of branched cyanides such as isopropyl cyanide (see Chapter 8). Both a branched carbon structure and chirality are common features in molecules that are needed for life, such as amino acids, which are the building blocks of proteins. Indeed, only a very small fraction of the organic compounds in Nature are found in planets or comets and other condensed objects. By far the larger quantity, more than 99.9% by mass, resides in the enormous molecular clouds in the interstellar space of the Milky Way and other galaxies. Thus, it seems natural to start from abiotic organic chemistry, as observed in molecular clouds, to catch a glimpse of the chemical evolution preceding the onset of life on our own planet, and to evaluate the possibility that – during the evolution from a molecular cloud to a planetary system – complex organic molecules are formed, transformed, and preserved until they are incorporated into comets, meteorites, and eventually planets.

What kind of complex molecules can we expect? As we have seen in previous chapters, the chemistry in interstellar regions is in many ways very different to the chemistry in terrestrial chemical laboratories that we are more accustomed to. Evidently, basic chemical concepts remain the same, but the physical conditions are such that some of the usual hypotheses are no longer valid. Through the development of exotic reaction methodologies to forge new chemical bonds, synthetic organic chemists are limited only by their imagination in constructing really complex organic molecules. Space chemistry, on the other hand, is not afforded such luxuries. The chemical processes in abiogenesis (*i.e.*, the origin of life from inanimate materials) had to occur under tight environmental constraints that a modern chemist undertaking routine synthesis in a laboratory would consider dreadful or archaic. Nevertheless, analyses of cosmic debris (such as meteorites) show that some amino acids and sugars contain significant excesses of enantiomers (molecules that are mirror images of each other; see Chapter 8) having the same handedness of terrestrial biomolecules. This coincidence is too striking to be fortuitous.

In this chapter we shall use some terms common in astrobiology. For convenience, some of these terms are listed in a glossary, see Box 10.1.

To conceive how life may begin on a habitable planet such as the Earth, it is essential to know what organic compounds were likely to have been available. The two key questions to be addressed are: how are simple organic compounds assembled into more complex molecular systems? And what are the essential processes and pathways by which complex systems can develop those basic properties – for instance homochirality – critical to the origin of life? From an astronomical perspective, chemistry is the most viable channel of investigation. Astrochemistry already deals, with noticeable success, with widely different conditions in elemental abundances, isotopic fractionation, and types of carbon bonds, showing that the organic compounds that make up life on Earth may possibly be the language of the Universe. We can then re-word our original questions as: to what extent is the origin of life linked to cosmic chemistry?

To answer to such a question, we need necessarily to start from the origins of our astronomical history. As we have seen in the previous chapter, planetary systems such as our own form from the collapse of an interstellar dense cloud composed of molecular gas and mutually colliding dust grains, eventually giving rise to a collection of bodies ranging from small particles to meteorites, and in a few tens of million years to planets. Beyond the snow line, dust grains accumulate mantles of ices

Box 10.1   A glossary of some terms used in astrobiology

Abiogenesis: the origin of life from inanimate materials

Abiotic: not derived from living material

Adenine: one of the four nucleobases in the nucleic acid of DNA (the others are guanine, cytosine, and thymine)

Amino acid: a molecule containing amine and carboxyl functional groups (see Box 8.1) and a specific side-chain

Anoxic: lacking in oxygen

Aqueous alteration: change in a rocky planet, asteroid or meteorite produced by interaction with water

Astronomical biosignature (biomarker): a molecular indicator of a biological process

Base: a base reacts with acids to form salts, accepts H^+ from donors, and contains OH^- ions

Base-pair: two nucleobases bound together by hydrogen bonds; the base-pairs form the building blocks of the DNA double-helix structure

Biogenic: formed in life processes

Carboxylic acid: an acid containing the –COOH group (see Box 8.1)

Chirality (a geometrical property): a molecule with a structure that cannot be super-imposed on its mirror image; these two structures are known as *enantiomers*

DNA (deoxyribonucleic acid): a molecule that carries the genetic instructions for the reproduction of all known organisms; it has the form of a double helix

Deoxyribose: a sugar that helps to form the backbone of the DNA molecule

Enantiomeric excess: the degree to which a sample contains more of one enantiomer than the other

Enantiomer: see Chirality

Endogenous: substances and processes originating from within an organism or cell

Great Oxidation Event: the plant-induced appearance of molecular oxygen in the Earth's atmosphere, which occurred around 2.5 billion years ago

Homochirality: uniformity of chirality

Hydroxy acid: a carboxylic acid with a substituted hydroxyl group

Late Heavy Bombardment: enhanced asteroid bombardment of the terrestrial planets around 4 billion years ago

Nucleic acid: inclusive name for DNA and RNA

Nucleobase: a component of nucleotides, the building blocks of nucleic acids

Nucleoside: a nucleotide without a phosphate group

Nucleotide: a unit of the structure of nucleic acids such as DNA and RNA

Oxidizing agent: a compound that causes other substances to lose electrons

Peptide: a chain of amino acids

Phosphoric acid: an organophosphorus compound containing the PO structure linked to three hydroxyl groups

Protein: a long chain of amino acid residues

Racemic mixture: a chiral molecule that has equal amounts of left- and right-handed enantiomers

(continued)

Box 10.1   *(continued)*

Reducing agent: a compound that loses an electron to another chemical
   species
Ribose: a sugar
RNA (ribonucleic acid): a chain of nucleotides often as a single strand folded
   onto itself, rather than the double-strand DNA
Sulfonic acid: an organosulfur compound containing the structure $R–S(=O)_2–$
   OH, where R is an alkyl (or other) group

---

containing a limited variety of fairly simple molecules, mainly
water, carbon monoxide, and carbon dioxide, with some other
simple molecules, such as ammonia. The chemical complexity
of these ices can be dramatically enhanced, through a solid-
state chemistry induced by environmental conditions, with the
more complex products eventually enriching the gas phase.
Because the formation of a planet is a violent event, the intricate
chemical history of the gas from which the planet forms may
be obliterated, requiring chemical evolution to be continuously
restarted. Nevertheless, the chemical mechanisms that generate biomolecules in space are replicated in proto-planetary systems, so that there is a potential connection between pre-biotic
organic chemistry and the ongoing chemistry we observe in the
interstellar medium.

Whichever way one looks at the possible origins of life, how
very complex molecules emerge from poorly organized chemical
systems, the conclusions are that life requires time, opportunity,
and much more complexity than that found in space. On the
other hand, organic species at the base of our biochemistry seem
to be natural products of cosmic chemistry, and their existence
is inextricably linked to dust. While we do not have a coherent
scenario for the origins of life on Earth, what we do know is that
cosmic dust has contributed to the emergence of life on Earth
many times during the intricate evolutionary pathway that eventually gave rise to our planetary system. Dust has contributed
precursors for pre-biotic chemistry in larger bodies, served as a
building block from which future comets, asteroids, and other
celestial bodies could originate, induced the formation of the
Earth itself and – more generally – planets, and delivered complex organics to the early Earth and Mars during the so-called
late heavy bombardment, about 4 billion years ago. The dust

chemical legacy might be further extended, as dust aggregates – resulting from the gentle gathering discussed in Section 9.5 – could provide stable and protected environments in which the pre-biotic synthesis of the building blocks of life can occur. Moreover, it could be that the dust grain structure favours left- or right-handedness, contributing thus to chiral selection.

Nature has had a four-billion-year head start in implementing controlled chemical evolution, and in doing that, a central role has been played by dust.

10.2 COSMIC CHEMICAL MESSENGERS

Those flashes of light romantically confused with shooting stars as they streak across the night sky are produced when *meteoroids* (see Box 2.2) pass through Earth's atmosphere. If a meteoroid survives a trip through the atmosphere and hits the ground, it's called a *meteorite*. Most meteoroids are tiny, and burn up as they fall toward Earth. The biggest ones may explode, as happened in 2013 in Russia, where a meteoroid as large as a house exploded above Chelyabinsk, a city of more than a million people.

Though meteorites may appear to be just exotic pieces of rock, they are extremely important, being left-overs from the formation of the Solar System. It is widely thought that they formed at the same time as the planets (see Chapter 9). While terrestrial rocks have been severely modified by geological forces over many eons, most meteorites have never experienced any re-processing and are just as they were when the Solar System was formed. In other words, they offer us the possibility to learn about the chemical composition of the Solar System as it was being born. Meteorites are either stony or metallic. A third class is constituted of fragments that come to us from the Moon or other planets, such as Mars. The most numerous meteorites are the stony meteorites. Some stony meteorites contain small, colourful, grain-like inclusions known as *chondrules*, rounded droplets of rocks that melted and then quickly cooled as they orbited the Sun. These tiny grains therefore pre-date the formation of our planet and the rest of the Solar System, making them the oldest known matter available to us for study. Stony meteorites that contain these chondrules are known as *chondrites*, and account for roughly 85% of all meteorite falls. Of all the chondrites, it is the *carbonaceous chondrites* that are considered the most primitive; these objects represent a window onto the earliest Solar System, showing no evidence of ever having been heated. They comprise about 3% of all meteorites collected after being seen to fall to Earth. These objects may contain up to 5% carbon, much of it organic. The carbon in carbonaceous chondrites is present in several forms, such as silicon carbide, graphite, nanometre-sized diamonds – originating as condensation products of stellar outflows as well as from catastrophic supernovae explosions that occurred long before the birth of the Sun – and as younger carbonate minerals. All of these forms of carbon are less abundant than organic matter. They can also contain up to 20% water by weight, and oxidized elements. With the highest proportion of volatile elements, they are thought to have originated from asteroids accreted at relatively large distances from the Sun.

A large body of detailed chemical analyses on carbonaceous chondrites found that the majority of organic matter (about 70–90%) is present as insoluble macromolecular material, something like terrestrial kerogen – the primary organic component of oil shale – consisting mainly of paraffin hydrocarbons (also called *alkanes*, see Box 8.1), major constituents of natural gas and petroleum. Organic chemists call these hydrogen-poor substances *tars*. The rest of the carbon is in a soluble complex mixture of compounds, in which the most notable (but not the most abundant) components are amino acids, nucleobases, and sugars, all of which are fundamental to life on Earth. There have also been identified other free molecules such as hydrocarbons, carboxylic acids, hydroxy acids, sulfonic acids, phosphonic acids, poly-hydroxyl compounds and many other chemical species that are of interest to the origins of life. The majority of the more than 80 different amino acids identified in carbonaceous meteorites are non-existent, or rare, in terrestrial proteins.

One would expect a similar molecular content in comets. Before the arrival of *Rosetta* near the comet 67P/Churyumov–Gerasimenko, our knowledge of cometary chemical composition was obtained by remote observations of volatile species, stored as ice in the cometary nucleus and released as gas into the cometary coma. The analysis of the gas by the instruments of the *Rosetta* orbiter and of its lander *Philae* has considerably improved the situation. All molecules viewed previously from the ground in 67P/Churyumov–Gerasimenko and other comets have been detected, plus some others. This chemical inventory is qualitatively and, in many cases, quantitatively consistent with that determined for astronomical ices as measured by infrared spectroscopy toward embedded proto-stars and background stars, confirming that this material is essentially unprocessed interstellar matter. Phosphorus, a key element in all known forms of life, has also been detected. Analysis of the free dust and of the surface of the nucleus of the comet shows the presence of silicates and organic materials. These materials, as well as molecules ejected by the comet, appear to contain some species relevant to biochemistry, including methyl isocyanate (CH_3NCO) and glycine (NH_2CH_2COOH), the simplest amino acid.

So far, glycine has been tentatively identified during an analysis of samples returned to Earth by NASA's STARDUST mission, which flew by Comet Wild 2 in 2004 to capture particles in its proximity. That glycine detection could not be confirmed because there were serious issues of contamination in terrestrial laboratories. From the Rosetta mission, however, glycine detection might be validated, because the mass spectrometer directly detected the mass of glycine, and there was no need for a chemical sample preparation that could have introduced contamination. However, this finding has been contested by a Japanese team, who pointed out that the reported detection is not yet convincing because the assignment is ambiguous, and there are other more likely candidates, such as glycolamide (NH_2COCH_2OH). The problem arises from the method used to identify molecular species, mass spectroscopy (see Box 8.3), that measures the mass of a compound. This technique (at least for the instrument on board Rosetta) is unable to distinguish between species with very similar masses. For instance, the masses of CO and N_2 molecules differ only by 0.02%. Thus, it is not possible to distinguish between them if the instrument is not sensitive enough. This does not mean that glycine is not there, simply that we cannot be sure! It's unfortunate, because the multitude of organic molecules identified

in 67P/Churyumov–Gerasimenko supports the idea that comets have the potential to deliver key molecules for pre-biotic
chemistry.

Thus far, amino acids have not been identified in the interstellar medium. However, two species containing the N–C=O bond
(perhaps the most important for proteins), isocyanic acid (HNCO,
the smallest stable molecule containing all four primary biogenic elements) and formamide (NH_2CHO) were detected in the
1970s, only a few years after radio astronomy opened the way
to the molecular Universe. The detection of two such species cannot be considered a fortuitous event. These rather simple molecules are in fact incredibly abundant in our Universe. Finding
amino acids, and in particular the simplest one, glycine, in interstellar space is considered the Holy Grail for astrobiologists.
Glycine is the only amino acid that is known to be able to form
without liquid water, and it could be synthesized within interstellar icy dust grains by ultraviolet irradiation, before becoming
bound up and conserved in a comet for billions of years. Two
independent studies by expert teams – one from the NASA Ames
laboratory in the USA and the other from Leiden Observatory
in Holland – found that amino acids were formed by 'zapping'
dirty water ices with ultraviolet radiation. The water ice suppos-
edly matched interstellar compositions and contained a fairly
high amount of ammonia, methanol and hydrogen cyanide. The
European researchers found 16 amino acids in the ice residue.
They repeated the experiments a second time replacing all of the
carbon atoms in the initial ice components with carbon-13 isotopes. Again 16 amino acids were found, and mass spectroscopy
clearly proved that all the carbon atoms of the generated amino
acids contained ^{13}C. Such results strongly suggest that amino
acids may be readily formed in interstellar space. So why we
don't see them out there? This is even more surprising when we
consider that the molecules that are potential precursors of glycine (aminoacetonitrile), or carry structural properties of amino
acids, such as the branched structure (isopropyl cyanide) and
chirality (propylene oxide) have been already detected in space.

But there is more. It's been known for years that scientists can
coax formamide to recombine into DNA base-pairs. Experiments
have shown that the production of base-pairs from formamide
can occur spontaneously following a cometary or asteroid

impact, or even collisions between two dust grains carrying icy mantles. Other experiments were designed to mimic chemical reactions that could naturally take place on a meteorite's surface. This was done simply by combining pure formamide with the ground-up powder of various meteorites in a cold environment, and processing them by simulating a stream of solar wind. In the products were found a wide-ranging list of more than 25 different molecules used by DNA and life in general, including nucleosides, structural sub-units of nucleic acids, the heredity controlling components of all living cells. Nucleosides consist of a molecule of sugar linked to a nitrogen-containing organic ring compound. In our biochemical machinery, the sugar is either ribose or deoxyribose, and the ring, one of the DNA and RNA bases.

Nevertheless, we have to keep in mind that the complexity of molecules in life is far greater than that seen in meteorites, and that the obstacles inherent to purely endogenous life origins would be the same as those faced by any developing organic system irrespective of the source of starting material. The minor bodies of the Solar System therefore probably present just a tiny and well-hidden sample of the pre-biotic potential of cosmic synthetic processes. All the matter in the Solar System bodies was once in the interstellar medium: how much of this matter survived initial incorporation into the proto-solar nebula and later the impact of a hot, energetic nebular chemistry? Transformations of organic compounds, or their synthesis from inorganics, occur in response to thermodynamic drivers, modulated by the kinetic properties of individual reactions, which in turn are entirely determined by the types and the ways the energy is deposited in the mixture, and above all by the physical environment in which the synthesis is taking place, which may channel and select among competing chemical outcomes.

Meteorites and asteroids preserve information about the early Solar System, and the processes active at those early times. One important process is *aqueous alteration*, the chemical reaction between co-accreted water and silicates, producing hydrated minerals. It has been suggested that amino acids and other organic compounds found in carbonaceous meteorites were formed by aqueous alteration in the meteorite parent bodies, *via* reactions of aldehydes and ketones with ammonia and HCN, in the so-called *Strecker synthesis* (see Box 10.2). Amino acids have also been found in meteorites that had experienced high temperatures. This suggests that the production of amino acids occurred through high-temperature processes (rather than the 'cold' Strecker synthesis) as their parent asteroids gradually cooled down. Such a process may involve gas containing hydrogen, carbon monoxide, and nitrogen, reacting in a Fischer–Tropsch scheme (see Box 10.2). This process is commonly used in industry to produce fuels (*i.e.*, complex hydrocarbons) by the catalytic hydrogenation of carbon monoxide.

These types of reactions are compatible with the existence of volatile ices in comets, and the chemistry associated with the formation and processing of interstellar ices, but it may be difficult to extend such an analogy to a higher degree of molecular complexity, as seen in meteorites. The role of interstellar processes, in particular those involving dust particles, if relevant, must then be examined from a different perspective.

10.3 THE SEARCH FOR THE PHYSICOCHEMICAL BASIS OF LIFE

Among the first laboratory experiments that addressed the problem of pre-biotic chemical evolution were those conducted at the University of Chicago by Stanley Miller and Harold Urey in the 1950s. Such experiments showed how the building blocks of the first self-replicating molecules could easily be made by cooking up a little primordial Earth atmosphere in a flask, by simply running a current of electricity through it (Figure 10.2). An electron in the current might collide with a molecule of gas and break its chemical bond, giving rise to something new. The flask held a mix of gases similar to those found in the Earth's early atmosphere – as Urey and Miller envisaged it – over a pool of water, representing Earth's early oceans. The electric current delivered by the electrodes into the gas-filled chamber simulated lightning storms that were considered to be a significant energy source in the Earth's primordial atmosphere.

Box 10.2   The Strecker synthesis and the Fischer–Tropsch process

Amino acids are organic compounds containing amine ($-NH_2$) and carboxyl ($-COOH$) functional groups, along with a side-chain (R group) specific to each amino acid. We describe here two synthetic processes that may work in the conditions of interstellar space: the Strecker synthesis and the Fischer–Tropsch process. In discussing these procedures we need to use the following chemical structures (see also Box 8.1):

- Aldehyde – a carbonyl group $>C=O$ bonded on one side to a hydrogen, and on the other side to a carbon that may belong to a functional group (*e.g.*, CH_3 in acetaldehyde);
- Imine – a modification of the carbonyl group in which nitrogen substitutes oxygen giving rise to a carbon–nitrogen double bond;
- Hydrocarbon – an organic compound consisting entirely of hydrogen and carbon; examples are methane, CH_4, that is also an alkane, and benzene, C_6H_6, the least complex aromatic hydrocarbon;
- Alkane – a hydrocarbon containing only carbon–hydrogen and carbon–carbon single bonds;
- Alkene – a hydrocarbon containing a carbon–carbon double bond;
- Alcohol – a carbon single bonded to an $-OH$ group;
- Amine – an organic derivative of ammonia, in which one, two, or all three of the hydrogens of ammonia are replaced by organic groups;
- Carboxyl group – an organic functional group, $-COOH$, consisting of a carbon atom double bonded to an oxygen and single bonded to a hydroxyl group ($-OH$);
- Acyl – the acyl group has the structure $R-C=O$ in a carboxylic acid or a carboxylic acid derivative; and
- Aminonitrile – a compound containing both an amino and a nitrile functional group; the simplest aminonitrile is cyanamide, H_2NCN.

The Strecker synthesis is a two-stage procedure used to synthesize amino acids. In Strecker's original experiment, the chosen aldehyde was acetaldehyde, which was made to react with NH_3 and HCN to produce the amino acid, alanine, $NH_2CH_3CHCOOH$. In the first stage, during the reaction of the acetaldehyde with ammonia, nitrogen 'exchanges' with oxygen, and the original carbonyl group becomes an imine (CH_3CHNH); the 'expelled' oxygen bonds to two hydrogens of the ammonia, forming water:

$$CH_3CHO + NH_3 \rightarrow CH_3CHNH + H_2O$$

Then HCN attacks the imine formed in the first step; HCN gives a proton (H^+) to the group $=NH$ of the imine, leading to the formation of the amino group $-NH_2$ (note that the double bond is now single, as we have two hydrogen atoms bonded). The remaining cyanide group, $-CN$, bonds to the central carbon forming an aminonitrile:

$$CH_3CHNH + HCN \rightarrow CH_3CH(NH_2)CN$$

(*continued*)

Box 10.2 (*continued*)

In the final stage, adding water to the aminonitrile, the –CN group is converted to a –COOH group, NH_3 is ejected and the amino acid alanine, $NH_2CH_3CHCOOH$ is formed.

Another mechanism leading to the formation of amino acids is the Fischer–Tropsch process. The general process is a catalytic reaction scheme in which a gas of molecular hydrogen and carbon monoxide is converted into a mixture of hydrocarbons, oxygenates, water and carbon dioxide. For more than half a century, this synthesis of liquid hydrocarbons was a technology of great potential for the indirect liquefaction of solid or gaseous carbon-based energy sources into liquid transportable fuels. The main products of the Fischer–Tropsch synthesis are olefins and paraffins, intermediate species for the production of sulfur-free diesel, gasoline and speciality chemicals. In general, low temperatures (*ca.* 250 °C) are applied for the production of long-chain paraffins while at higher temperatures (*ca.* 350 °C) lighter products are obtained.

A crucial aspect of the Fischer–Tropsch reaction is that it is a highly exothermic reaction, *i.e.*, it does not require an energy input as in the Urey–Miller synthesis, where the mixture has to be heated and 'zapped' with electricity to mimic lightning. The Fischer–Tropsch reaction occurs on catalytic surfaces (for instance, a meteoritic surface), and it is essentially a hydrogenation reaction of carbon monoxide. The main chemical reactions in a Fischer–Tropsch synthesis are the production of:

- Alkanes – $(2n+1)H_2 + nCO \rightarrow C_nH_{2n+2} + nH_2O$
- Alkenes – $2nH_2 + nCO \rightarrow C_nH_{2n} + nH_2O$
- Alcohols – $2nH_2 + nCO \rightarrow C_nH_{2n+1} + (n-1)H_2O$

Since ammonia may be adsorbed on surfaces and can dissociate in adsorbed N, N–H or N–H_2 surface species, the formation of nitrogen-containing compounds such as amines is feasible under Fischer–Tropsch conditions. This might go through the hydrogenation of surface *acyl* intermediates to form a compound in which the central carbon is bound (in addition to the surface) to a hydrogen, a hydroxyl group, and a generic organic functional group R. As shown in Figure 10.1 the hydroxyl group is replaced by an adsorbed amino species –NH_2 on the catalyst surface, then the loss of a water molecule results in the formation of an amine. In the example shown in the figure, if R is the carboxyl group, –COOH, the reaction produces glycine. Water in liquid form is considered a critical ingredient for life. However, with Fischer–Tropsch reactions, all the ingredients needed are hydrogen, carbon monoxide, and nitrogen as gases – which are all very common molecules in space – and suitable surfaces. Thus, chemistry can begin making some pre-biotic components of life very early, before asteroids or planets with liquid water are formed.

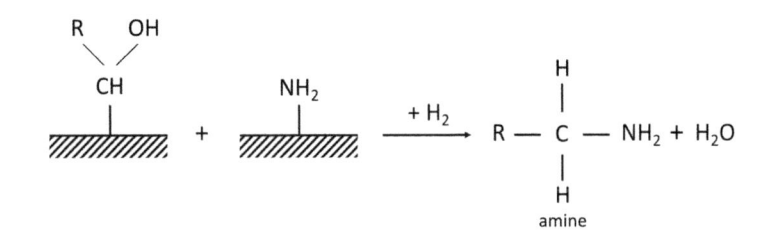

Figure 10.1   Fischer–Tropsch scheme for the formation of amines. The basic Fischer–Tropsch process converts carbon monoxide and hydrogen to hydrocarbons. However, in the presence of nitrogen hydrides, amines may also be formed, as indicated schematically in this diagram. R represents an organic functional group.

Since the 1950s, chemists have been drawn toward studies of the origin of life. Melvin Calvin, an American biochemist at the University of California and his associates began pioneering experiments in pre-biotic evolution by irradiating a solution of carbon dioxide and water vapour with ionizing radiation from the Crocker Laboratory's 60-inch cyclotron. These experiments had limited success, leading to the production of only formic acid and formaldehyde. The Urey–Miller experiment carried out three years later showed how compounds of biochemical importance could be produced in high yields from a mixture of hydrogen-rich gases. The experiment established that the early Earth atmosphere was capable of producing amino acids, the building blocks of life, from inorganic substances. It was very impressive that the Urey–Miller results were published only a few weeks after Watson and Crick's classic article describing the double-helix model for the structure of DNA.

In the 1950s, the generally accepted model for the evolution of the atmosphere considered the early Earth's atmosphere to be anoxic, *i.e.*, an environment without free oxygen gas (O_2), very different from the one we breathe today. In fact, the Urey–Miller experiments were driven by Urey's theory that Earth accreted as a cool body and that its atmosphere was dominated by hydrogen and the hydrides of common volatiles. For decades scientists believed that the atmosphere of early Earth was highly hydrogen-rich, or with a greatly limited content of oxygen. Such conditions would have resulted in a primordial atmosphere filled with nitrogen (N_2), methane (CH_4), water vapour (H_2O), and possibly

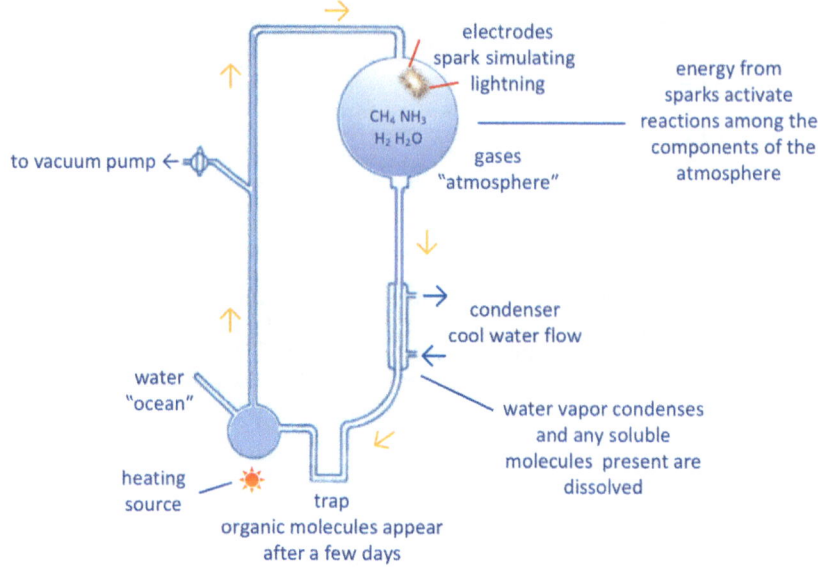

Figure 10.2 A schematic view of the Urey–Miller experiment. The diagram indicates the various components that are active in the experiment, and the location of the trap where organic molecules appear after a few days of operation.

ammonia (NH_3). A reducing environment such as that would tend to donate electrons to the atmosphere, leading to reactions that form more complex molecules from simpler ones. Today's oxidizing atmosphere, in which oxygen is present, works in the opposite way, stripping electrons from chemical bonds, so that pre-biotic molecules would be destroyed as fast as they could be produced. It would be very difficult to produce such molecules in the presence of an oxidizing atmosphere. The reducing conditions, as well as high energy levels from UV radiation and strong lightning and volcanic discharges on primitive Earth, were thus believed to have set the stage for the origin of life. The reducing-atmosphere hypothesis was very important to understanding the origin of organic compounds on primitive Earth, and confirmed the lack of spontaneous generation in an oxidizing environment, as observed in the experiments of Louis Pasteur, who proved that bacteria could not spontaneously grow in a filtered broth under oxidizing conditions. The Urey–Miller experiment impacted dramatically on the study of the origins of life, which acquired a novel, quantitative perspective. In 1957 Miller showed that formaldehyde and hydrogen cyanide were key intermediates in the synthesis of glycine. This led Joan Oró and his co-workers to study, a few years later, the products of a solution of ammonium cyanide (NH_4CN) in water, discovering that NH_4CN was converted to adenine, one of the four bases of DNA. Such (and other similar) discoveries determined the direction of research on pre-biotic chemistry for many years.

From the late 1960s onward, however, it became clear that our understanding of the origins of life was troubled by two major issues. Firstly, the possibility that the young Earth had been characterized by a highly reducing atmosphere was viewed with increasing scepticism by most planetary scientists, who favoured mostly CO_2-rich atmospheres, weakening the possibility of an abiotic synthesis of organic molecules (much more efficient in hydrogen-rich atmospheres). Secondly, we know that DNA depends heavily on proteins for functioning, and at the same time DNA holds the recipe for protein construction. This DNA–protein paradox appeared to be an insurmountable problem. A possible solution to such a chicken-and-egg conundrum would be the evolution of a molecule that could copy itself. The RNA World hypothesis is based on the discovery that RNA could both carry information and cause chemical reactions like those required to copy a molecule (see Box 10.3). The RNA molecule as a whole is however too complex, and the difficulties inherent in its creation are not very far from those involved in DNA production. Each of its building blocks is chemically complex, and biochemists are still exploring whether these species could be formed by chance.

A common perception (or misconception) of pre-biotic chemistry is that Nature has an innate tendency to produce life's building blocks preferentially, rather than the plethora of compounds that could be derived from the rules of organic chemistry. From the results of the Urey–Miller experiments to findings in meteorites, inanimate Nature appears to be generous in providing a supply of, for instance, amino acids and sugars. Then, it could be presumed that all of life's building blocks could be formed with ease in Miller-type experiments and found in meteorites and other extra-terrestrial bodies. That's not the case, as thorough investigations of meteoritic material have shown: abiotic

Box 10.3   The RNA World

The RNA World hypothesis is a concept put forward in the 1960s by Carl Woese, Francis Crick and Leslie Orgel, to solve the chicken-and-egg conundrum posed by the structure of growth shared by all living organisms. Walter Gilbert, a Harvard molecular biologist, was the first to use the term "RNA World" in an article published in 1986.

The genetic information flows from DNA to RNA to proteins. DNA encodes RNA, which directs the synthesis of proteins. Proteins do most of the biochemical work in cells and are required for the structure, function, and regulation of the body's tissues and organs, and to capture energy. This energy is directed into the synthesis of new copies of DNA, resulting in the growth of new cells and organisms. Proteins are particularly efficient in performing all these tasks because they speed up enormously the reactions in biological systems by lowering the activation barrier needed to start that reaction. It is the protein's shape (as encoded in the RNA) that determines its function.

How could proteins get made without the DNA code for them? And without proteins, how could DNA (and all the rest of the biochemical machinery) be created in the first place? In other words, proteins require DNA to store information, but DNA also requires proteins to do biochemical work. Neither DNA nor protein can support life alone. Their mutual relation is, by far, too complex to have emerged spontaneously. Something simpler must have preceded it. The RNA World states that RNA was the first to play the double role of storing information and catalyzing chemical reactions, opening the way for DNA and proteins to take over later.

Structurally, DNA and RNA are nearly identical. They are polymers composed of monomers called *nucleotides*. While DNA consists of two strands, arranged in a double helix, RNA has only one strand. DNA and RNA perform different functions. DNA is responsible for storing and transferring genetic information, while RNA directly codes for amino acids and acts as a messenger. The RNA World hypothesis states that there was a primordial soup and, in this soup, there existed free nucleotides. These nucleotides could form bonds with one another. These chains have been proposed by some as the first, primitive forms of life. In the initial stage, RNA molecules performed the catalytic activities necessary to assemble themselves from a nucleotide soup. Through trial and error, the RNA molecules evolved in self-replicating patterns, developing an entire range of enzymic activities. At the next stage, RNA molecules began to synthesize proteins, initially as adaptor molecules, *i.e.*, acting as a 'bridge' between other molecules, allowing information to be transmitted. Then, they acted as a template using other RNA molecules such as the RNA core of the ribosome. Since proteins are much better enzymes than their RNA counterparts, they substituted for RNA in this job. Finally, DNA appeared on the scene, being a better holder of the information copied from the genetic RNA molecules by reverse transcription, relegating RNA to the intermediate role it has today.

The RNA World has become a widely accepted hypothesis for the origin of life. That's mainly because the RNA World is fortified by much more experimental evidence than any of its competitors have accumulated. However, the RNA World hypothesis remains a subject of fractious debate, with some claiming that the theory does not provide a sufficient foundation for the evolutionary events that followed. One of the main objections to RNA concerns catalysis, and it has been shown that for life to take hold, the unknown pristine polymer would have had to coordinate the rates of chemical reactions that could differ in speed by as much as 20 orders of magnitude. Perhaps most importantly, it has recently been evidenced that an RNA-only World could not explain the emergence of genetic code, because it is just plain chemistry and totally lacks what Charles Parker and Peter Wills, biophysicists studying the origin of genetic coding, call *reflexive information*: information that, "when decoded by the system, makes the components that perform exactly that particular decoding". Evidently, the last chapter hasn't been written yet, and the story is not over.

synthesis fosters the formation of molecules made of fewer rather than greater numbers of carbon atoms. As we have seen before, when larger carbon-containing molecules are produced, they tend to be insoluble, hydrogen-poor substances called tars. On the other hand, it would be rather over-simplifying to consider the reactivity of the simplest pre-biotic feedstock molecules as giving rise to uncontrollable reaction pathways that produce complex mixtures containing a multitude of undesirable products and intractable tars.

The essential components of the biochemical machinery, such as amino acids and nucleobases, are far less complex than nucleotides and proteins. We might then expect that the building blocks of biological molecules could be synthesized in space, and then assembled on a planet in much more complex chemical entities. Life on our planet could have been started somewhere between 3.7 and 4.5 billion years ago, after meteorites and dust splashed down and leached essential elements into a small basin of water, that were then bonded together as water levels fell and rose through cycles of precipitation, evaporation and drainage. This scenario has been suggested in a recent work by Ben Pearce and colleagues at McMaster University in Canada. Life would begin when the Earth was still taking shape, with continents emerging from the oceans, meteorites pelting the planet, and no protective ozone to filter the Sun's ultraviolet rays. It's intriguing that this vision revives Charles Darwin's old idea of the origin of life in a warm little pond.

There are hints of life in very primeval rocks. A rock in Western Australia, dated to be 4.1 billion years old, contains high amounts of a form of carbon typically used in biological processes. In 2017, putative microfossils were discovered in sedimentary rocks from the Nuvvuagittuq belt in Quebec, Canada. The fossils are at least 3.8 million and possibly 4.3 million years old, suggesting an "almost instantaneous emergence of life" (at least in the geological sense), after the ocean formation occurred 4.4 billion years ago. The earliest direct evidence of life on Earth is of microorganisms mineralized in a 3.5 billion year old microcrystalline silica deposit in Australia. Earth is estimated to be about 4.5 billion years old, and thus for much of its history it has been home to life. Was biological evolution on Earth jump-started by a special delivery from outer space? Those chemicals wouldn't have been abundant on a brand-new planet, which had a very different and much simpler chemical composition than today's Earth.

10.4   CHEMICAL MICROREACTORS IN SPACE

The physical conditions and chemical reactions of the kind Stanley Miller and others envisaged and explored in early prebiotic experiments may not be characteristic of only planetary environments. We have seen that circumstellar discs are a natural consequence of the star-formation process. These discs can be considered proto-planetary because of the geometry of the Solar System and the high detection rate of exoplanets. Once formed, proto-planetary discs evolve, forming planets, satellites, asteroids, and comets before they dissipate within a few tens of millions of years. During the evolution of these discs, molecules freeze-out from the gas phase onto dust grain surfaces while dust grains grow through collisional agglomeration. This process ends up in the formation of loosely packed structures with much of their internal volume being vacuum and trapped ices. Interstitial voids must occur even in highly organized, densely packed structures. For example, dense packing of spheres in a face-centred cubic lattice leaves 26% of space unoccupied (Figure 10.3). It is this internal volume that leads to a completely different chemical scenario with respect to surface chemistry occurring in the icy grain mantles.

As we saw in Chapter 7, surface chemistry proceeds in a sequence of steps. First of all, a future reactant must collide with the grain surface, and then accommodate and stick. This species may react at the site of impact, or undergo a delayed reaction with

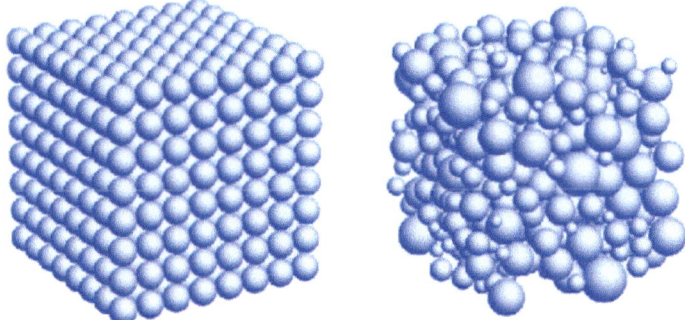

Figure 10.3   A cubic assembly of spheres of equal size is shown in the left-hand image. In this structure 26% of space is unoccupied. The right-hand image shows a similar cubic assembly for spheres of a range of sizes.

other grain adsorbates. The time delay between the adsorption and reaction steps may be substantial if the reaction products are non-volatile, but can be short for prompt reactions. The products of surface reactions are either retained on dust or desorbed to the ambient gas. In 2000, Walter Duley pointed out that the interiors of dust aggregates, those internal voids produced by the accretion process, offer a different intermediate possibility: the re-accretion of reaction products onto other components of the aggregate. As desorbed products can be in an energetic state, these secondary reactions might mimic some aspects of high-temperature chemistry.

Dust aggregates can also be impulsively heated by collisions with other aggregates or grains and by cosmic ray impacts. Cosmic rays possess an energy that is extremely high, much higher than normal thermal energies. The interaction of cosmic rays with dust gives rise to so-called *sputtering*. This is a process whereby atoms of the grain materials (such as Si, Ca, Mg, Fe, and P) are dislodged and ejected from the surface. The heat released during a collision may lead to the vaporization of the ice content filling the cavities. As a consequence, part of the chemical species forming the ice enter a transient, warm, high-pressure gas phase, together with sputtered atoms from the grain substrate, in a hydrogen-rich atmosphere. The cavities within the grain aggregate bring together all the components of gas and dust, a unique situation outside planetary systems. The resulting mixture is a reasonable analogue of the conditions that Urey and Miller envisaged as plausible for the primitive Earth's atmosphere. In this respect, grain aggregates represent the equivalent of terrestrial microlaboratories containing raw materials of reducing chemical composition suitable for conversion into more complex organic species. In other words, the Urey–Miller experiment is revisited through innumerable repetitions inside dust grain aggregates. The chemical products contained in these aggregates would then be incorporated into planetesimals and comets that may fall on a planet, leading to a plausible connection between the chemistry in cold interstellar regions and that in material falling onto planets.

We know that the early Earth had been subjected to the constant threat of impacts with the left-overs of planet-building material. Such a violent activity occurred 4.1 to 3.8 billion years ago. Close to the end of this period, impacts in the Solar System may have increased, because of the migration of giant planets, which sent debris into scattered inner orbits that intercepted those of smaller rocky worlds. The evidence for two early-bombardment populations is now emerging, and that there was a time difference between them. The late one came plausibly from the Asteroid Belt, while the early one from elsewhere. This first episode might have been caused by failed young planets or planetesimals – thought to be far larger than the objects in the Asteroid Belt – that would have done significant damage as they crashed into the rocky inner Solar System planets. During this horrific era, it is thought that most of Earth's water had been vaporized, and perhaps the exterior of our planet had been sterilized, erasing any life form that might have succeeded in emerging. Thus, prebiotic chemistry had to wait for the storm to pass to take firm root and eventually give rise to all later life. Even if cometary and asteroidal impacts had destructive effects, refreshing the chemical conditions to raw materials both in their interiors and on the primordial Earth, the gentle dusty rain of smaller dust particles provided a continuous supply of organic chemically organized materials. If life did originate on the Earth's surface, then it either could have happened many times before the end of the major impact era, or, if it started only after the conditions became less hostile, it succeeded very quickly. In either hypothesis, grain aggregates are an ideal factory, providing a stable and reducing environment for the ingredients needed to start life early and quickly.

Except in rare cases, life uses only the left-handed form of amino acids, while in the Miller experiment these species are racemic, meaning that they are synthesized in mixtures with equal numbers of left- and right-handed forms. This is not necessarily a critical flaw: both could form, but one enantiomer may be more easily destroyed than the other. In the next section we discuss how the presence of cavities in dust aggregates might also favour the emergence of a specific handedness.

10.5   THE ROLE OF DUST IN CHIRAL SELECTION IN SPACE

As we discussed in Chapter 9, a chiral molecule exists in two forms called *enantiomers*, the mirror images of each other. Individual enantiomers are often designated as either left-handed or right-handed. In living organisms, proteins are made uniquely of left-handed amino acids, while RNA and DNA are made only of right-handed sugars. The reason for this behaviour, called *homochirality*, is so far an unfilled gap in the theories of the chemical origin of life. The two enantiomers have the same chemical reactivity, the same mass, volume, and so on. However, they can be easily distinguished: enantiomers react differently with other chiral molecules, and they are *optically active*.

This activity arises in two important phenomena: *optical rotation* and *circular dichroism*. As we have seen in Box 1.1, light waves are electromagnetic waves, which can move in any direction. If light waves have their electric fields vibrating in a fixed plane, perpendicular to the direction of propagation, then the light is said to be *plane* or *linearly polarized* (see Figure 1.6). Identically prepared solutions of the two separate enantiomers of a chiral molecule rotate such a polarization plane (Figure 10.4, left panel) in equal but opposite directions. A racemic mixture of chiral molecules does not produce such an effect. Circular dichroism – only observed for chiral substances – is related to *circularly polarized light*, meaning that the light wave is the result of the super-position of two waves

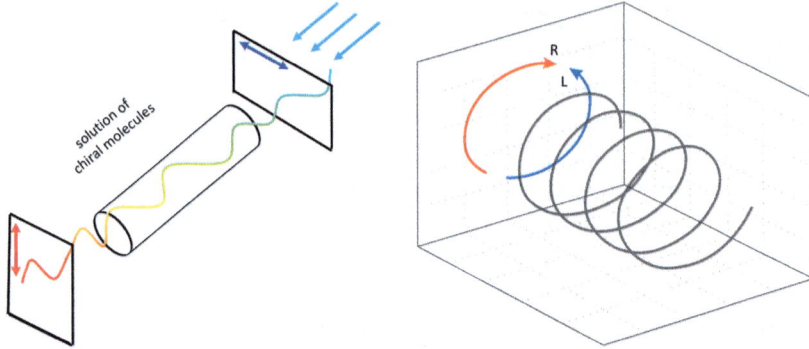

Figure 10.4 Plane and circular polarization of light. (Left) Indicates how the plane of polarization of a beam of light may be changed from its original state (shown by the blue arrow) by passage through a solution of chiral molecules to a new state (as shown by the red arrow). (Right) Indicates the motion of a particle riding a circularly polarized wave. If the motion as seen facing the propagating wave occurs in a counter-clockwise way, the wave is said to be circularly *left-polarized*, otherwise it is *right-polarized*.

that are linearly polarized in mutually perpendicular planes. If we were able to place a particle on the wave, we would see it oscillating in a straight line if the wave was linearly polarized, or moving in a corkscrew-like fashion if the wave was circularly polarized [Figure 10.4 (right)]. In this latter case, looking along the direction of the incoming wave, the particle may rotate in either counter-clockwise or clockwise ways, defining *right-handed* and *left-handed* polarized waves, respectively. To some extent we may say that circularly polarized light is chiral in itself. In fact, any helical object has a mirror image that is also helical, but with the opposite sense of rotation. Thus, it is not surprising that the interaction of circularly polarized light of one handedness may be different with the two enantiomers of a chiral molecule. If the absorption spectrum of one enantiomer is measured using circularly polarized light, the results will be slightly different if it is recorded with left- or right-hand circularly polarized light. The difference between the two absorption spectra is called the *circular dichroism spectrum*, and would have the opposite sign if we had taken, as absorber, the other enantiomer.

Several controversial theories have been developed to explain an abiogenic origin of the chiral homogeneity. Some have argued that equal numbers of both versions of each chiral molecule were present at the onset of life and that it was only during biological evolution that the imbalance occurred. That view has become increasingly unpopular, however, with the realization that the fundamentally important process of protein folding seems to require chiral imbalances, while for Nature to have selected the left- or right-handedness of each molecule during evolution would involve extraordinarily complex processes. Since the discovery by Michael Engel and Bartholomew Nagy in 1982 of an enantiomeric excess of left-handed amino acids in meteorites (and more recently of right-handed sugars), it has been evident that it is possible to create asymmetrical molecules in space conditions from a mixture that does not initially contain any chiral substances.

We have seen that different enantiomers exhibit differential absorption of left and right circularly polarized light. If the absorbed radiation has a destructive effect – as generally occurs for ultraviolet radiation – the initial equality of enantiomers in a racemic mixture will change, with the concentration of the more strongly absorbing enantiomer decreasing. This results in a measurable imbalance called *enantiomeric excess*. Most proposed extra-terrestrial mechanisms involve ultraviolet circularly polarized light from some astronomical source acting on chiral molecules in the molecular cloud from which the Solar System formed, to produce an excess of one enantiomer. Unfortunately, ultraviolet circularly polarized light is a rare phenomenon in space. Observations of young stellar objects reveal the existence of circularly polarized light in the infrared range, irrelevant to the aim of molecular destruction. Radiation from white dwarfs can be highly circularly polarized but any effect on molecular clouds and star-formation regions must rely on rare chance encounters. Diffuse starlight in the ultraviolet region is either unpolarized or partially linearly polarized through dichroic extinction by interstellar grains (see Section 1.4). It is this linearly polarized component that offers a way to solve the problem: when traversing a grain aggregate, in the intricate network of connected voids, different linearly polarized waves tend to add together acquiring a degree of circularity. This effective ultraviolet circular polarization is generated *in situ*, and exposes the amino acids that can be formed in the cavities of dust aggregates, described in the previous section, to asymmetric destruction.

However, the effect is purely geometric and in this sense is not chiral, since it does not provide symmetry breaking. In fact, for symmetry reasons, the sign of the induced circular polarization changes with the rotation of the cavities with respect to the propagation of the wave. A possible way to break the symmetry of the process is provided by the same mechanism that acts to bring dust grains to clump together, the *gas drag*. During their motion in the rotating disc, dust grains try to minimize drag, and in doing that the aggregates can be efficiently aligned, with the denser part leading and the more porous one following. The resulting motion is similar to that of a badminton shuttlecock. In fact, the 'birdie' has a conical shape and, because the cork is much denser than the feathers, a non-homogeneous mass. Once the aggregate is aligned, because the more porous part has more cavities, a net enantiomeric excess of chiral molecules can be established. At the same time, the drag – the same force that makes it harder to flip – coaxes the grains into clusters, which then rapidly collapse into solidity *via* their own gravity, forming larger chunks (see Section 9.5). The chemicals stored inside grain aggregates would be then more and more protected from the extremely unfavourable environmental conditions at the beginning of the Sun's life.

In conclusion, grain aggregates provide the perfect environment in which interstellar molecules can evolve into complex organic species. The final step of conversion of these molecules into biopolymers is dependent on their delivery to planets offering favourable conditions for the life experiment to become possible. We know that – at least on the Earth – this experiment was successful!

10.6 THE SEARCH FOR LIFE IN OTHER PLANETARY SYSTEMS

The flourishing activity of organic chemistry in space is even more significant as we realize that our Milky Way galaxy is literally teeming with exoplanets. Statistically speaking, there is at least one planet for each of the hundreds of billions of stars in the galaxy. In fact, thousands of planets have been discovered, with thousands more planet candidates identified. Organic chemistry in space and the emergence of life on Earth suggest that such a synthesis may have occurred elsewhere in the Solar System and the countless extra-solar worlds.

The examination of living organisms surviving in extreme environments (so-called *extremophiles*) on Earth is considerably expanding our understanding of the limits of life, going closer to the conditions in which life may have arisen. The very existence of extremophiles provides us with a wondrous array of life's adaptations, making the search for life outside the Earth more plausible. Life may be plentiful. The only way to find out is to search. Still, when we find life, how will we know? The search for life in the Universe requires consideration of how we can detect signs of life elsewhere, and intelligent life in particular, and more specifically what constitutes evidence for life? This gives rise to the notion of an *astronomical biosignature* (or *biomarker*), that may unfortunately be rather misleading because our concepts of life and biosignatures are inextricably linked. This opens the door to a more difficult question: *just what, exactly, is life?*, which has been occupying humankind for centuries. While scientists have proposed hundreds of ways to define it, none have been widely accepted. The problem with most proposed definitions of life is that they have loopholes, and the answer to such a question is still open-ended. Biosignatures must reflect fundamental and universal characteristics of life, and they should not be restricted solely to those attributes that represent local solutions to the challenges of survival. For instance, certain specific mechanisms of our biosphere, such as DNA and proteins, might not necessarily be mimicked by other examples of life elsewhere in the cosmos. On the other hand, organic species at the base of our biochemistry seem to be natural products of cosmic chemistry. Thus, basic evolutionary principles might indeed be universal.

Finding any life that might exist on other planets is extremely challenging. Even in the Solar System, where we can send probes and orbiters to worlds of interest such as Mars, it is rather difficult to assess if any microbial life is, or was ever, present. Astronomers are beginning to probe the upper atmospheres of some exoplanets. The starlight shining through a planet's atmosphere reveals absorption or emission lines of the constituent gases, and among them those produced by life, such as oxygen (O_2) and methane (CH_4). Much of the search for biosignatures has focused on O_2, whose production on Earth is now primarily by photosynthesis. But while, so far, oxygen remains central to the search for biosignatures, there are some serious problems with relying on that molecule, because O_2 can also be produced abiotically, and therefore by itself is not a conclusive biosignature. In fact, over the last decades, there has been increasing evidence of the presence of oxygen on Mars. Carbon dioxide-rich atmospheres can form oxygen by the break-up of CO_2. This suggests that some oxygen must have been created in the early Earth's atmosphere as well, due to the similar compositions of the two atmospheres. Before, it was widely understood that oxygen in the Earth's atmosphere originated in an event called the *Great Oxidation Event*, which occurred about 2.4 billion years ago as the first plants appeared and converted carbon dioxide to oxygen. Life perturbs disequilibria that arise due to kinetic barriers, and can give unexpected structure to an abiotic system. What would be more discriminating is a mixture of gases in disequilibrium that can only be maintained by biotic and not abiotic processes. Abiotic processes, unless continually sustained, tend towards equilibrium. For example, on Earth, if life completely disappeared today, our nitrogen–oxygen dominated atmosphere would eventually reach equilibrium with the oxygen bound as nitrates in the ocean.

Looking at the crescent Moon shortly after sunset or before sunrise, you may have noticed a dull glow that lights up the unlit part of the Moon. That pale glow is light reflected from Earth. It's called *earthshine*. Earthshine allow us to observe the Earth as a distant planet, without spatial resolution. Earthshine happens when the light from the Sun is reflected twice, once off the Earth's surface and then off the Moon's surface. This transmission spectrum carries with it information on the Earth's surface. In fact, astronomers have identified in the spectrum key signatures of life, finding clear signs of water and an oxygen atmosphere. Their findings give a strong indication of what 'fingerprints' we should search for when seeking life on Earth-like worlds orbiting distant stars. Among these data, they also found features that suggested the presence of chlorophyll, indicating the existence of land plants, in the form of bright reflections in the far-red region of the visible spectrum. This red edge is a well-known signature of chlorophyll, which appears green to us only because our eyes aren't very sensitive at the red end of the visible spectrum. The Earth's atmosphere has included a significant component of oxygen for the past 2 billion years, and must have shown a 'red edge' since the first land plants evolved 500 million years ago. This strong reflectance of Earth's vegetation suggests that surface biosignatures with sharp spectral features might be detectable in the spectrum of scattered light from a spatially unresolved extra-solar terrestrial planet.

In addition to biosignatures, NASA is searching for alien life using *techno-signatures* that may emanate from advanced civilizations. These techno-signatures have generally been limited to communication signals, but any kind of evidence, such as radio or laser emissions, signs of massive structures or an atmosphere full of pollutants could imply intelligence. However, according to the Fermi paradox which states that "if another intelligent life form was indeed out there, we would have met it by now", it is not clear whether we'll ever find evidence of advanced civilizations. The oddity in the behaviour of a bizarre star, known to astronomers as KIC 8462852 – more popularly called Tabby's Star for its discoverer Tabetha Boyajian – whose flux dims by a tremendous amount, without any regularly repeating signals, has brought scientists to suggest a number of possible objects that could be blocking the star's light, from a family of large comets to even "alien megastructures" orbiting the star. Such alien structure, named a *Dyson sphere* after the physicist Freeman Dyson who first explored this idea as a thought experiment in 1960, would be a highly advanced piece of technology able to intercept the power of the star and divert it back to a planet. In an article published in the journal *Science*, Dyson speculated the existence of stellar power collection systems, with sizes comparable to planetary orbits that advanced civilizations in our galaxy could have constructed. Dyson went further and proposed to search for evidence of such structures, as proof of the existence of advanced civilizations elsewhere in the galaxy.

The existence of a Dyson sphere is an extremely fascinating idea. But thanks to a huge number of follow-up observations, we know that it's wrong. The observations have been performed in an extended spectral range, from blue light all the way to infrared light, with the blue light preferentially blocked in all dimming events. We know what causes bluer light to be blocked while redder light is preferentially transmitted: the obscuring material is quite ironically, actually simple dust, drifting between us and the star in tendrils of varying thickness. The source of the dust, however, remains a mystery. In any case, even if alien megastructures exist somewhere, Tabby's Star is not where we should look.

10.7   SUMMARY: WHAT WE KNOW AND WHAT WE NEED TO KNOW

In this chapter we have put together a remarkable story that links interstellar chemistry with star and planet formation to suggest that the origins of life on Earth – and perhaps elsewhere – may lie in space. It's an appealing narrative, made up of things we know and understand with speculations founded on intuition, together with some steps based on imagination rather than science. It's a complicated situation, and this culminating chapter has been one of the more demanding chapters in this book. So it's worth standing back for a moment and trying to summarize the situation.

The starting point for our story is straightforward. The interstellar medium is rich in molecules, and especially in organic molecules, and – as we have seen in earlier chapters of this book – dust grains play a crucial role in much of this chemistry. It's only in the last half-century that we've begun to realize that the interstellar medium can be not only molecular, but that interstellar chemistry is remarkably complex for what seems a quite hostile environment. Of course, the chemistry of life is much more complex than interstellar chemistry. But wherever life begins, it must start from the kinds of molecules that are readily available in interstellar space. So our narrative has a perfectly logical foundation.

What are these molecules? More than 200 different molecular species have been identified in interstellar space, and most of these are organic (carbon-containing) molecules. These molecules are caught up in the processes that make stars and planets, sticking to dust grains that combine to make meteoroids, planetesimals, asteroids, comets, and eventually planets. Objects like meteoroids and comets carry with them samples of the products of interstellar chemistry that we can examine, and these samples are rich in molecules that are necessary to make the building blocks of DNA and RNA. In particular, meteorites are rich in amino acids, nucleobases, and sugars. So, the next stage of our narrative is well founded.

It's not obvious how these essential molecules can be made in an interstellar or planet-forming environment and transported safely to a planet. However, it seems likely that interstices in clusters of dust grains can provide environments that may mimic the Urey–Miller experiment in which a variety of molecules, including amino acids and sugars, were created in abundance. Perhaps clusters of dust grains could be the mechanism of forming and transporting essential molecules to planet Earth. Clusters may also provide environments in which it is possible to create an enantiomeric excess in optically active molecules. So this next stage is somewhat uncertain but very encouraging.

However, there are many things that we don't understand about this narrative. For example:

- Why has the amino acid glycine (for example) not yet been identified in the interstellar medium? Species that are chemically close to glycine have been observed.
- The formation mechanisms of many interstellar complex molecules are still obscure. Does surface chemistry on interstellar dust grains play an important role in making most of these species? Is there a limit to the chemical complexity produced by these routes?
- The idea of Fischer–Tropsch type processing of molecular species trapped inside grain clusters is very compelling. Experimental studies are required. Do these processes chemically select certain types of products?

- Can dust clusters promote chiral selection, as described above?
- Our discussion goes only as far as the provision of the basic building blocks of DNA and RNA. The next steps are still unclear. Do they proceed in the RNA World? How can we explore the mechanisms of assembly of the building blocks?

CHAPTER 11

What Have We Learned About Dust in Space?

11.1   THE COMPLETE NARRATIVE OF DUST IN SPACE

The modern, comprehensive story about dust in space runs like this:

- Dust grains are currently present in galaxies throughout that part of the Universe that we can observe, and have been present in the Universe for more than 98% of the time that the Universe has existed.
- Dust grains have been recognised as a component of the Milky Way and other galaxies by the patchy fog that they create in interstellar space; this fog causes a partial or total extinction of starlight.
- This extinction exists from the infrared to the ultraviolet, and confirms the need for dust grains of a range of sizes comparable to wavelengths over this range. Polarization of starlight by partially aligned asymmetric dust grains emphasizes the existence of dust grains comparable in size to the wavelength of visible light.

Dust in Galaxies
By David A. Williams and Cesare Cecchi-Pestellini
© David A. Williams and Cesare Cecchi-Pestellini 2020
Published by the Royal Society of Chemistry, www.rsc.org

- Dust grains are formed in the extended envelopes of cool stars and in supernovae and novae ejecta, and ejected into interstellar space.
- In the interstellar gas, dust grains evolve by irradiation with starlight, by accretion of atoms and molecules, and by atom–grain and grain–grain collisions in high-velocity shocks. Erosion and shattering of dust grains in these collisions ensure that there are many more small grains than large.
- By a variety of inferences from observations, and from the evidence of collected interstellar grains, it is clear that bare interstellar dust grains are composed mainly of solid silicates and solid carbons, with small PAHs and carbon nanoparticles also present.
- Bare dust grains in the Milky Way catalyze the formation of molecular hydrogen, which is essential to enable an extensive chemistry to occur in interstellar gas.
- Dust grains shield simple molecules – such as carbon monoxide, water, and hydroxyls – from starlight and allow their abundances in interstellar clouds to rise to significant levels. These molecules radiate energy from low-density interstellar clouds, sustaining the collapse of these clouds to denser, darker states; this is the initial stage of star formation.
- In dark clouds, dust grains provide surfaces on which simple molecular ices are deposited; these simple ices are the feedstock for the formation of complex organic molecules in dense clouds.
- Dust grains permit the formation of stars in interstellar dark clouds not only through the provision of simple molecular coolants that help to keep the dark cloud cool while the collapse occurs, but also by the radiation emitted from the grains themselves during the collapse. The collapse continues until the newly-forming star begins to heat itself and its surroundings.
- Dust grains are the sources of the materials that make up planets, comets, asteroids, and meteoroids, which are formed in the proto-planetary disc as an accompaniment to star formation. Specifically, the atoms contained in everything on planet Earth have come originally from interstellar dust.

- Complex molecules formed from simple ices on dust grains, and in clusters of dust grains, can be transported within the proto-planetary disc and deposited on newly-forming planets.
- These complex molecules are expected to be rich in amino acids, nucleobases and sugars, and may help to trigger the formation of DNA and RNA and the evolution of life on planets.
- Dust in proto-planetary discs may provide geometrical environments in which one form of a chiral molecule is preferentially selected, facilitating terrestrial homochirality.

It really is a very powerful narrative, showing that dust grains have a number of absolutely fundamental roles to play in the Universe as we know it today. Different parts of the story may be emphasized by one story-teller or another (for example, the roles of dust in chemistry or in star formation) but it is the comprehensive picture of dust throughout the whole saga that makes it compelling. This *complete* story makes clear that for a Universe like ours, dust is required to make molecular hydrogen and initiate chemistry, including complex chemistry, to permit the formation of stars like those we can see. Dust also provides the raw material for planets, and seeds them with the molecular building blocks of life. These building blocks are formed on dust in the same star- and planet-forming regions, as those shown in Figure 11.1. We can ask: will places like these proto-planetary discs host our future neighbours?

11.2 ARE OTHER TECHNOLOGICAL CIVILIZATIONS PRESENT IN THE MILKY WAY?

Of course, dust tells us a great deal about chemistry, the formation of stars, planets and the possible origins of life, but says absolutely nothing about whether life does evolve on a planet (or even elsewhere), and – if so – what kind of life it might be. Interesting speculations on these lines have often been approached using the so-called *Drake Equation*, written down by Frank Drake in 1961 to identify the various factors that need to be considered in attempting to estimate the number

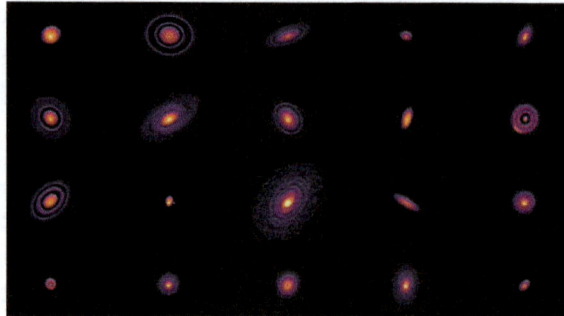

Figure 11.1   ALMA images of 20 nearby proto-planetary discs. These striking false-colour infrared images show the variety of structures
found in nearby proto-planetary discs. It seems evident that all
of these discs are in the process of forming planets. The implied
rate of planet formation in the Milky Way galaxy appears to be
high [credit: ALMA (ESO/NAOJ/NRAO), S. Andrews *et al.*; NRAO/
AUI/NSF, S. Dagnello].

of *technological civilizations* that exist in the Milky Way galaxy.
These civilizations are assumed to be at least as technically
capable as the human race on planet Earth, but are in fact more
likely to be technically far in advance of us, and so capable of
communicating in various ways with aspirations or abilities in
interstellar travel. The Drake Equation (don't worry, it's the only
equation in this book) is written in this form:

$$N = R \times f_p \times n_e \times f_l \times f_i \times f_c \times L$$

where N is the number of technological civilizations in the
Milky Way; this is the number that we seek to find. The sym-
bols on the right-hand side of the equation are simply factors
that may (or may not) affect the actual number N. The equation
assumes that if a technological civilization is to emerge, it will
do so on a planet (which is a reasonable but possibly incorrect
assumption).

The first factor on the right-hand side of the equation, R_*, is
the average rate of star formation in the Milky Way galaxy, per
year, and the second factor, f_p, is the fraction of those stars that
have planets, while the third factor, n_e, is the number of planets
forming in the galaxy that are capable of hosting life – at least,
life as we understand it. After all, planets may be too hot, too
cold, too fluid, too dry, too irradiated, *etc.*, for life to emerge. So,

multiplying the first three factors on the right-hand side of the equation (if we know them!) should tell us the number of planets being formed in the Milky Way each year capable of hosting life.

The next factor, f_l, is the fraction of all those planets that actually do develop life at some point, and the factor f_i is the fraction of all those planets on which intelligent life develops – after all, it's no good trying to communicate with the green scum on a dirty pond. Not all civilizations will release (or choose to release) detectable signals into space, so we need a factor f_c, which is the fraction of all these civilizations that do give signs of their existence to the galaxy. So, multiplying all these factors together, the equation should tell us the rate at which planets arriving in the Milky Way each year are actively sending signals out into the galaxy.

Finally, L is the lifetime in years of such civilizations. Multiplying the rate of formation of technically active civilizations, per year, by the average life time of such civilizations, in years, should give us the number that we seek: the average number of technological civilizations in the galaxy with which we might be able to communicate.

Astronomical observations have developed enormously since Drake first proposed these original ideas, and we know much more about star formation and planet formation in the galaxy than we did 60 years ago. We know that the star-formation rate in the galaxy is on average about one star per year. However, the greatest advance has been in the detection of planets orbiting other stars. We now know that planets are common and that some at least are Earth-like and may therefore be suitable for life as we know it. Indeed, one such exoplanet has recently (2019) been identified. We also know much more about our own history on Earth: human existence on Earth may have lasted for about 200 000 years, civilization in cities for about 12 000 years, and technological capability (so far) merely for a century or so, but it is almost unbounded, in principle.

If we take an optimistic case and say that all stars have planets and at least one planet per star is habitable for life, that all such planets will develop life and that life will ultimately be intelligent and eventually technologically capable, then (with these assumptions) the number of technological civilizations in the galaxy is numerically equal to L. Who can say, from a sample of one civilization (*i.e.*, ours), how long technological civilizations might survive? One would like to think human civilization might survive for many thousands – or even millions – of years, in which case there may be as many as a hundred thousand technological civilizations in the Milky Way, or even many more. That's a prospect that some find exhilarating, others find terrifying. The idea is the stuff of science fiction.

But the great Italian physicist Enrico Fermi asked this simple question: if there really are many technological civilizations in the Milky Way, many of which will be technologically far in advance of human civilization on Earth, why have we not been visited? Does this mean that the number of civilizations in the Milky Way is very small, possibly only one – *i.e.*, ours? What a responsibility that would be!

Others have said: "No, there are probably many civilizations in the galaxy". These external civilizations may have Earth under observation, but will not contact us until we are culturally and emotionally ready to receive such a signal. In other words, could it be the case that civilizations like ours are being held in a 'zoo' of immature technological civilizations? Are we being perceived as not ready to join the 'grown ups'?

11.3   WHAT IF FERMI'S ANSWER IS CORRECT?

Perhaps the simplest answer to Fermi's question would be that we haven't been visited because there aren't many technological civilizations; perhaps there's only one – ours. The reason may be that such civilizations do not survive for very long. With increasing technical ability comes the power of destruction in war and the ability to despoil the home planet by neglect or by deliberate act, as in human-induced climate change on Earth. Both war and environmental neglect could lead to the demise of Earthly technological civilization as we know it.

We are beginning to understand the origin of stars and planets, and taking our first steps to understand the origin of life on planet Earth; the role of interstellar dust is surprisingly wide and important in all these areas. In principle, the lifetime, L, of our civilization on Earth need be bounded only by the life of our Sun. What a tragedy it would be if our own actions limited the size of L to a number that is infinitesimally small compared to those billions of years that could potentially be available to us.

11.4 CONCLUSION: THE UNIVERSE AS A MACHINE FOR PRODUCING LIFE?

In this book we have presented a coherent narrative that implies that dust grains have active and essential functions in the Universe in which we live. Those functions are to drive chemistry in interstellar space, to contribute to the formation of planetary systems around new stars, to provide complex pre-biotic molecules and deliver them to young planets, and thereby to promote the origin of life on planets.

Of course, if dust isn't present, then these functions cannot be fulfilled. In the early Universe – before elements such as carbon, nitrogen, and oxygen were formed in the first generation of stars – dust grains did not exist. The dust grains only became significant players after that very first generation of star formation began, and the Universe gradually became similar to the one we observe today.

The broad sweep of the narrative is convincing, but of course there is much more work to do. For example, much of the excellent and convincing experimental and theoretical work on which the narrative is based depends on a much simpler picture of interstellar dust and gas than can exist in space. Our understanding of the formation of complex organic molecules in space is still rudimentary. The role of Fischer–Tropsch chemistry in dust grain clusters has yet to be explored.

Nevertheless, we know enough to be sure that the picture is broadly true. It leads to the remarkable conclusion that the Universe could be regarded as "a machine for producing life" and that dust is an essential component in that machine.

Subject Index